The HACKER *Ethic*

Prologue

LINUS TORVALDS

Epilogue

MANUEL CASTELLS

RANDOM HOUSE / NEW YORK

The HACKER *Ethic*

and the Spirit of the Information Age

PEKKA HIMANEN

Pekka Himanen's text translated by Anselm Hollo and Pekka Himanen

Published in the United States by Random House, Inc., New York, and simultaneously in Canada by Random House of Canada Limited, Toronto

Grateful acknowledgment is made to the following for permission to reprint copyrighted material:

Richard Stallman: "The Free Software Song," by Richard Stallman. Copyright © 1993 by Richard Stallman. Used by permission. Verbatim redistribution permitted if this notice is preserved.

Youth Radio, Berkeley, California: E-mails between Finnegan Hamill and "Adona." Copyright © Youth Radio, Berkeley, California. Used by permission.

Library of Congress Cataloging-in-Publication Data
Himanen, Pekka.
The hacker ethic, and the spirit of the information age / Pekka Himanen.
p. cm.
Includes bibliographical references and index.
ISBN 0-375-50566-0
1. Computer programming—Moral and ethical aspects. 2. Computer hackers.
3. Open source software. I. Title.

QA76.9.M65 H56 2001
174'.90904—dc21 00-053354

Random House website address: www.atrandom.com

Printed in the United States of America on acid-free paper

2 4 6 8 9 7 5 3

First Edition

Contents

PART THREE: THE NETHIC

CONCLUSION

Preface

At the core of our technological time stands a fascinating group of people who call themselves *hackers.* They are not TV celebrities with wide name recognition, but everyone knows their achievements, which form a large part of our new, emerging society's technological basis: the Internet and the Web (which together can be called the Net), the personal computer, and an important portion of the software used for running them. The hackers' "jargon file," compiled collectively on the Net, defines them as people who "program enthusiastically"[1] and who believe that "information-sharing is a powerful positive good, and that it is an ethical duty of hackers to share their expertise by writing free software and facilitating access to information and to computing resources wherever possible."[2] This has been *the hacker ethic* ever since a group of MIT's passionate programmers started calling themselves hackers in the early sixties.[3] (Later, in the mid-eighties, the media started applying the term to computer criminals. In order to avoid the confusion with virus writers and intruders into information systems, hackers began calling these destructive computer users *crackers*.[4] In this book, this distinction between hackers and crackers is observed.)

My own initial interest in these hackers was technological, related to the impressive fact that the best-known symbols of our time—the Net, the personal computer, and software such as the Linux operating system—were actually developed not by enterprises or governments but were created primarily by some enthusiastic individuals who just started to realize their ideas with other like-minded individuals working in a free rhythm. (Those who are interested in the details of their development may turn to the appendix, "A Brief History of Computer Hackerism," for details of their development.) I wanted to understand the internal logic of this activity, its driving forces. However, the more I thought about computer hackers, the more obvious it became that what was even more interesting about them, in human terms, was the fact that these hackers represented a much larger spiritual challenge to our time. Computer hackers themselves have always admitted this wider applicability of their ways. Their "jargon file" emphasizes that a hacker is basically "an expert or enthusiast of any kind. One might be an astronomy hacker, for example."[5] In this sense, a person can be a hacker without having anything to do with computers.

The main question transformed into, What if we look at hackers from a wider perspective? What does their challenge then mean? Looking at the hacker ethic in this way, it becomes a name for a general passionate relationship to work that is developing in our information age. From this perspective, the hacker ethic is a new *work ethic* that challenges the attitude toward work that has held us in its thrall for so long, *the Protestant work ethic,* as explicated in Max Weber's classic *The Protestant Ethic and the Spirit of Capitalism* (1904–1905).[6]

To some computer hackers, this kind of linking of the hacker ethic to Weber may at first seem alien. They should keep in mind that in this book the expression *hacker ethic* is used in a sense that extends beyond computer hackerism, and that for this reason it confronts social forces that are not normally considered in discussions concerned exclusively with computers. This expansion of the hacker ethic thus presents an intellectual challenge to computer hackers, as well.

But first and foremost the hacker ethic is a challenge to our society and to each of our lives. Besides the work ethic, the second important level of this challenge is the hacker *money ethic*—a level that Weber defined as the other main component of the Protestant ethic. Clearly, the "information-sharing" mentioned in the hacker-ethic definition cited above is not the dominant way of making money in our time; on the contrary, money is mostly made by information-owning. Neither is the first hackers' ethos—that activity should be motivated primarily not by money but rather by a desire to create something that one's peer community would find valuable—a common attitude. While we cannot claim that all present computer hackers share this money ethic or that it is likely to spread into society at large, as we can about their work ethic, we can say that it has been an important force in the formation of our time and that the hackers' debate over the nature of the information economy could lead to consequences at least as radical as those of their work ethic.

The third element present within the hacker ethic from the very beginning, touched upon in the cited definition by the phrase "facilitating access to information and to computing resources," could be called their network ethic or *nethic.* It has addressed ideas such as freedom of expression on the Net and access to the Net for all. Most computer hackers support only some parts of this nethic, but in terms of their social significance they must be understood as a whole. The impact of these themes remains to be seen, but they definitely go to the heart of the ethical challenges of the information age.

This book is based on an ongoing collaboration between its three authors, a collaboration taking place in various forms over several years (with Manuel Castells through research we conduct together in California, and with Linus Torvalds in the midst of having fun). The idea for a book dealing with the hacker ethic was born the first time all three of us met, in the fall of 1998, when we were invited speakers at a symposium hosted by the University of California at Berkeley, that traditional hacker stronghold. At that time, we decided to expand our presentations, which dealt with the same subjects as the present work. Linus, we decided, would start as a representative of computer hackerism, Manuel would present his theory of our information age (consisting of the rise of informationalism, the new information-technology paradigm, and a new social form, the network society), and I would examine the social meaning of the hacker ethic by placing the example of Linus's computer hackerism against Manuel's larger background picture of our time. Naturally, each one of us would still speak for himself.

The book adheres to this plan: in his Prologue, "What Makes Hackers Tick? a.k.a. Linus's Law," Linus—as the originator of one of the most famous hacker creations of our time, the Linux operating system—describes his view of the forces that contribute to the success of hackerism. Manuel has spent the last fifteen years on a study of our time, culminating in his three-volume, 1,500-page work, *The Information Age* (second revised edition, 2000). In this book's epilogue, "Informationalism and the Network Society," he presents for the first time the findings of his research, with some new important additions, in a form accessible to the general reader. My analysis is placed between Linus's and Manuel's and is divided into three parts according to the three levels of the hacker ethic: the work ethic, the money ethic, and the nethic. (Some further elaborations of these themes can be found at the book's website, www.hackerethic.org.)

Those readers who prefer to have a description of the theory background before, and not as a closing systematization of, my examination, may consult Manuel's epilogue right away. Otherwise, let Linus start.

PROLOGUE

What Makes Hackers Tick? a.k.a. Linus's Law

LINUS TORVALDS

I first met with Pekka and Manuel at an event that the University of California at Berkeley had put together in the Bay Area, a half-day symposium on the challenges of the network society. Here were these social-science big shots talking about modern technology and society. And there I was, representing the technical side.

Now, I'm not an easily intimidated person, but this wasn't exactly the kind of setting in which I was most comfortable. How would my opinions fit in with those of a bunch of sociologists talking about technology? But hey, I thought, if they have sociologists talking about technology, they might as well have a technologist talking about sociology. At worst, they'd never invite me back. What did I have to lose?

I always end up doing my talks the day before, and once

again, there I was feverishly trying to get an "angle" for the next day. Once you have that angle—your platform—writing a few slides is usually not that hard. I just needed an idea.

I ended up setting out to explain what makes hackers tick and why Linux, the small operating system that I started, seems to appeal so strongly to hackers and their values. In fact, I ended up reckoning, not just with hackers, but with our highest motives in general. I called my notion (in my normal humble and self-deprecating way) "Linus's Law."

Linus's Law

Linus's Law says that all of our motivations fall into three basic categories. More important, progress is about going through those very same things as "phases" in a process of evolution, a matter of passing from one category to the next. The categories, in order, are "survival," "social life," and "entertainment."

The first phase, survival, is a truism. Any living thing needs to survive as its first order of business.

But the other two? Assuming you agree with *survival* as being a fairly fundamental motivational force, the others follow from the question "What are people ready to die for?" I'd say that anything for which you might forfeit your life has to be a fairly fundamental motivation.

You can argue about my choices, but I think they work.

You can certainly find instances of people and other living creatures who value their *social ties* more than they do their lives. In literature, *Romeo and Juliet* is the classic example, of course, but you can also think about the notion of "dying for your family/country/religion" as a way of explaining the notion of social ties as potentially more important than life itself.

Entertainment may sound like a strange choice, but I mean by *entertainment* more than just playing games on your Nintendo. It's chess. It's painting. It's the mental gymnastics involved in trying to explain the universe. Einstein wasn't motivated by survival when he was thinking about physics. Nor was it probably very social. It was entertainment to him. Entertainment is something intrinsically interesting and challenging.

And the quest for entertainment is certainly a strong urge. You might not feel the urge to die for your Nintendo, but think of the expression "dying of boredom": some people, certainly, would rather die than be bored forever, which is why you find people jumping out of perfectly good airplanes—just for the thrill of it, to keep boredom at bay.

What about money as a motivation? Money is certainly useful, but most people would agree that money per se is not what ultimately motivates people. Money is motivational for what it brings—it's the ultimate bartering tool for the things we *really* care about.

One thing to note about money is that it's usually easy

to buy survival, but it is much harder to buy social ties and entertainment. Especially Entertainment with a capital *E*—the kind that gives your life meaning. One should not dismiss the social impact of having money, whether you buy something or not. Money remains a powerful thing, but still it is just a proxy for other more fundamental motivating factors.

Linus's Law per se is not so much concerned with the fact that these three things motivate people but more with the fact that our progress is a matter of going through the full phase change from "survival" to "social life" to "entertainment."

Sex? Sure. It obviously started out as survival, and it still is. No question about that. But in the most highly developed animals, it's progressed past being a thing of pure survival—sex has become part of the social fabric. And for human beings, the pinnacle of sex is entertainment.

Eating and drinking? Check. War? Check. Maybe war is not quite there yet, but CNN is doing its best to get it to that final stage. It certainly started out as survival, has progressed to a means of maintaining social order, and is inexorably on its way to becoming entertainment.

Hackers

All this definitely applies to hackers. To hackers, survival is not the main thing. They'll survive quite well on Twinkies and Jolt Cola. Seriously, by the time you have a computer on your desk, it's not likely that your first worry is how to get the next meal or keep a roof over your head. Survival is still there as a motivational factor, but it's not really an everyday concern to the exclusion of other motivations anymore.

A "hacker" is a person who has gone past using his computer for survival ("I bring home the bread by programming") to the next two stages. He (or, in theory but all too seldom in practice, she) uses the computer for his social ties—e-mail and the Net are great ways to have a community. But to the hacker a computer is also entertainment. Not the games, not the pretty pictures on the Net. The computer itself is entertainment.

That is how something like Linux comes about. You don't worry about making that much money. The reason that Linux hackers do something is that they find it to be very interesting, and they like to share this interesting thing with others. Suddenly, you get both entertainment from the fact that you are doing something interesting, and you also get the social part. This is how you have this fundamental Linux networking effect where you have a lot of hackers working together because they enjoy what they do.

Hackers believe that there is no higher stage of motivation than that. And *that* belief has a powerful effect in realms far beyond that of Linux, as Pekka will demonstrate.

PART ONE

The WORK *Ethic*

CHAPTER 1

The Hacker Work Ethic

Linus Torvalds says in his Prologue that, for the hacker, "the computer itself is entertainment," meaning that the hacker programs because he finds programming intrinsically interesting, exciting, and joyous.

The spirit behind other hackers' creations is very similar to this. Torvalds is not alone in describing his work with statements like "Linux hackers do something because they find it to be very interesting." For example, Vinton Cerf, who is sometimes called "the father of the Internet," comments on the fascination programming exerts: "There was something amazingly enticing about programming."[1] Steve Wozniak, the person who built the first real personal computer, says forthrightly about his discovery of the wonders of programming: "It was just the most intriguing world."[2] This is a general spirit: hackers program because programming challenges are of intrinsic *in-*

terest to them. Problems related to programming arouse genuine curiosity in the hacker and make him eager to learn more.

The hacker is also *enthusiastic* about this interesting thing; it energizes him. From the MIT of the sixties onward, the classic hacker has emerged from sleep in the early afternoon to start programming with enthusiasm and has continued his efforts, deeply immersed in coding, into the wee hours of the morning. A good example of this is the way sixteen-year-old Irish hacker Sarah Flannery describes her work on the so-called Cayley-Purser encryption algorithm: "I had a great feeling of excitement. . . . I worked constantly for whole days on end, and it was exhilarating. There were times when I never wanted to stop."[3]

Hacker activity is also *joyful*. It often has its roots in playful explorations. Torvalds has described, in messages on the Net, how Linux began to expand from small experiments with the computer he had just acquired. In the same messages, he has explained his motivation for developing Linux by simply stating that "it was/is fun working on it."[4] Tim Berners-Lee, the man behind the Web, also describes how this creation began with experiments in linking what he called "play programs."[5] Wozniak relates how many characteristics of the Apple computer "came from a game, and the fun features that were built in were only to do one pet project, which was to program . . . [a game called] Breakout and show it off at the club."[6]

Flannery comments on how her work on the development of encryption technology evolved in the alternation between library study of theorems and the practice of exploratory programming: "With a particularly interesting theorem . . . I'd write a program to generate examples. . . . Whenever I programmed something I'd end up playing around for hours rather than getting back to plodding my way through the paper."[7]

Sometimes this joyfulness shows in the hacker's "flesh life" as well. For example, Sandy Lerner is known not only for being one of the hackers behind the Internet routers but also for riding naked on horseback. Richard Stallman, the bearded and longhaired hacker guru, attends computer gatherings in a robe, and he exorcises commercial programs from the machines brought to him by his followers. Eric Raymond, a well-known defender of hacker culture, is also known for his playful lifestyle: a fan of live role-playing games, he roams the streets of his Pennsylvania hometown and the surrounding woods attired as an ancient sage, a Roman senator, or a seventeenth-century cavalier.

Raymond has also given a good summary of the general hacker spirit in his description of the Unix hackers' philosophy:

> To do the Unix philosophy right, you have to be loyal to excellence. You have to believe that software is a craft worth all the intelligence and passion you

> can muster. . . . Software design and implementation should be a joyous art, and a kind of high-level play. If this attitude seems preposterous or vaguely embarrassing to you, stop and think; ask yourself what you've forgotten. Why do you design software instead of doing something else to make money or pass the time? You must have thought software was worthy of your passions once. . . .
>
> To do the Unix philosophy right, you need to have (or recover) that attitude. You need to *care.* You need to *play.* You need to be willing to *explore.*[8]

In summing up hacker activity's spirit, Raymond uses the word *passion,* which corresponds to Torvalds's *entertainment,* as he defined it in the Prologue. But Raymond's term is perhaps even more apt because, even though both words have associations that are not meant in this context, *passion* conveys more intuitively than *entertainment* the three levels described above—the dedication to an activity that is intrinsically interesting, inspiring, and joyous.

This passionate relationship to work is not an attitude found only among computer hackers. For example, the academic world can be seen as its much older predecessor. The attitude of passionate intellectual inquiry received similar expression nearly 2,500 years ago when Plato, founder of the first academy, said of philosophy, "Like light flashing forth when a fire is kindled, it is born in the soul and straightway nourishes itself."[9]

The same attitude may also be found in any number of other spheres of life—among artists, artisans, and the "information professionals," from managers and engineers to media workers and designers, for example. It is not only the hackers' "jargon file" that emphasizes this general idea of being a hacker. At the first Hacker Conference in San Francisco in 1984, Burrell Smith, the hacker behind Apple's Macintosh computer, defined the term as follows: "Hackers can do almost anything and be a hacker. You can be a hacker carpenter. It's not necessarily high tech. I think it has to do with craftsmanship and caring about what you're doing."[10] Raymond notes in his guide "How to Become a Hacker" that "there are people who apply the hacker attitude to other things [than software], like electronics and music—actually, you can find it at the highest levels of any science or art."[11]

Looked at on this level, computer hackers can be understood as an excellent example of a more general work ethic—which we can give the name *the hacker work ethic*—gaining ground in our network society, in which the role of information professionals is expanding. But although we use a label coined by computer hackers to express this attitude, it is important to note that we could talk about it even without any reference to computer people. We are discussing a general social challenge that calls into question the Protestant work ethic that has long governed our lives and still maintains a powerful hold on us.

Let's see what type of long historical and strong societal

forces the hacker work ethic, in this sense, faces. The familiar expression "Protestant work ethic" derives, of course, from Max Weber's famous essay *The Protestant Ethic and the Spirit of Capitalism* (1904–1905).[12] Weber starts out by describing how the notion of work as a duty lies at the core of the capitalist spirit that arose in the sixteenth century: "This peculiar idea, so familiar to us today, but in reality so little a matter of course, of one's duty in a calling, is what is most characteristic of the social ethic of capitalistic culture, and is in a sense the fundamental basis of it. It is an obligation which the individual is supposed to feel and does feel towards the content of his professional activity, no matter in what it consists, in particular no matter whether it appears on the surface as a utilization of his personal powers, or only of his material possessions (as capital)." Weber goes on to say: "Not only is a developed sense of responsibility absolutely indispensable, but in general also an attitude which, at least during working hours, is freed from continual calculations of how the customary wage may be earned with a maximum of comfort and a minimum of exertion. Labour must, on the contrary, be performed as if it were an absolute end in itself, a calling."[13]

Then Weber demonstrates how the other main force described in his essay, the work ethic taught by Protestants, which also arose in the sixteenth century, furthered these goals. The Protestant preacher Richard Baxter expressed that work ethic in its pure form: "It is for action that God maintaineth us and our activities; work is the moral as well as the natural end of power," and to say "I will pray and meditate [instead of working], is as if your servant should refuse his greatest work and tie himself to some lesser, easier part."[14] God is not pleased to see people just meditating and praying—he wants them to do their job.

True to the capitalist spirit, Baxter advises employers to reinforce this idea in workers of wanting to do one's job as well as possible by making it a matter of conscience: "A truly godly servant will do all your service in obedience to God, as if God Himself had bid him do it."[15] Baxter sums up this attitude by referring to labor as a "calling,"[16] a good expression of the three core attitudes of the Protestant work ethic: work must be seen as an end in itself, at work one must do one's part as well as possible, and work must be regarded as a duty, which must be done because it must be done.

While the hacker work ethic's precursor is in the academy, Weber says that the Protestant ethic's only historical precursor is in the monastery. And certainly, if we expand on Weber's comparison, we can see many similarities. In the sixth century, for example, Benedict's monastic rule required all monks to see the work assigned to them as their duty and warned work-shy brethren by noting that "idleness is the enemy of the soul."[17] Monks were also not supposed to question the jobs they were given. Benedict's fifth-century predecessor John Cassian made this clear in his monastic rule by describing in admiring tones the obedience of a monk, named John, to his elder's order to roll a stone so large that no human being could move it:

> Again, when some others were anxious to be edified by the example of his [John's] obedience, the elder called him and said: "John, run and roll that stone hither as quickly as possible;" and he forthwith, applying now his neck, and now his whole body, tried with all his might and main to roll an enormous stone which a great crowd of men would not be able to move, so that not only were his clothes saturated with sweat from his limbs, but the stone itself was wetted by his neck; in this too never weighing the impossibility of the command and deed, out of reverence for the old man and the unfeigned simplicity of his service, as he believed implicitly that the old man could not command him to do anything vain or without reason.[18]

This Sisyphean straining epitomizes the idea, central to monastic thought, that one should not question the nature of one's work.[19] Benedict's monastic rule even explained that the nature of the work did not matter because the highest purpose of work was not actually to get something done but to *humble* the worker's soul by making him do whatever he is told—a principle that seems to be still active in a great number of offices.[20] In medieval times, this prototype for the Protestant work ethic existed only within the monasteries, and it did not influence the prevailing attitude of the church, much less that of society at large. It was only the Protestant Reformation that spread the monastic thinking to the world beyond the monastery walls.

However, Weber went on to emphasize that even though the spirit of capitalism found its essentially religious justification in the Protestant ethic, the latter soon emancipated itself from religion and began to operate according to its own laws. To use Weber's famous metaphor, it turned into a religiously neutral iron cage.[21] This is an essential qualification. In our globalizing world, we should think of the term *Protestant ethic* in the same way we think of an expression such as *platonic love.* When we say that someone loves another person platonically, we do not mean that he is a Platonist—that is, an adherent of Plato's philosophy, metaphysics and all. We may attribute a platonic love relationship to a follower of any philosophy, religion, or culture. In the same way, we can speak of someone's "Protestant ethic" regardless of his or her faith or culture. Thus, a Japanese person, an atheist, or a devout Catholic may act—and often does act—in accordance with a Protestant ethic.

One need not look very far to realize how strong a force this Protestant ethic still is. Commonplace remarks like "I want to do my job well," or those made by employers in their little speeches at employee retirement parties about how a person "has always been an industrious/responsible/reliable/loyal worker" are the legacy of the Protestant ethic in that they make no demands on the nature of the work itself. The elevation of work to the status of the most important thing in life—at its extreme, a work addiction that leads to complete neglect of one's loved ones—is another symptom of the Protestant ethic. So is work done with clenched jaws and a responsibility-ridden attitude and the bad conscience many feel when they have to miss work due to ill health.

Seen in a larger historical context, this continued dominance of the Protestant ethic is not so surprising when we remember that even though our network society differs in many significant ways from its predecessor, the industrial society, its "new economy" does not involve a total break with the capitalism Weber describes; it is merely *a new kind of capitalism.* In *The Information Age,* Castells stresses that work, in the sense of labor, is not about to end, despite wild paradisiacal forecasts such as Jeremy Rifkin's *The End of Work.* We easily fall for this illusion that technological advances will, somehow, automatically, make our lives less work-centered—but if we just look at the empirical facts of the rise of the network society so far and project them into the future, we must agree with Castells on the nature of the prevailing pattern: "Work is, and will be for the foreseeable future, the nucleus of people's life."[22] The network society itself does not question the Protestant ethic. Left to its own devices, the work-centered spirit easily continues to dominate within it.

Seen in this overall context, the radical nature of general hackerism consists of its proposing an alternative spirit for the network society—a spirit that finally questions the dominant Protestant ethic. In this context, we find the only sense in which all hackers are really crackers: they are trying to crack the lock of the iron cage.

The Purpose of Life

The displacement of the Protestant ethic will not happen overnight. It will take time, like all great cultural changes. The Protestant ethic is so deeply embedded in our present consciousness that it is often thought of as if it were just "human nature." Of course, it is not. Even a brief look at pre-Protestant attitudes toward work provides a healthy reminder of that fact. Both the Protestant and the hacker ethic are historically singular.

Richard Baxter's view of work was completely alien to the pre-Protestant church. Before the Reformation, clerics tended to devote time to questions such as "Is there life after death?" but none of them worried about whether there was *work* after life. Work did not belong among the church's highest ideals. God himself worked for six days and finally rested on the seventh. This was the highest goal for human beings as well: in Heaven, just as on Sundays, people would not have to work. Paradise was in, office was out. One might say that Christianity's original answer to the question "What is the purpose of life?" was: the purpose of life is Sunday.

This statement is not just a witticism. In the fifth century, Augustine compared our life quite literally to Friday, the day when, according to the teachings of the church,

Adam and Eve sinned and Christ suffered on the cross.[23] Augustine wrote that in Heaven we'll find a perennial Sunday, the day on which God rested and Christ ascended to Heaven: "That will truly be the greatest of Sabbaths; a Sabbath that has no evening." Life is just a long wait for the weekend.

Because the Church Fathers saw work as merely a consequence of the fall from grace, they also took very particular conceptual care in their descriptions of Adam's and Eve's activities in Paradise. Whatever Adam and Eve may have done there, it could not be seen as *work.* Augustine emphasizes that in Eden "praiseworthy work was not toilsome"—it was no more than a pleasant *hobby.*[24]

The pre-Protestant churchmen understood work, "toil," as punishment. In medieval visionary literature that speaks to churchmen's images of Hell, the implements of labor fully reveal their true nature as instruments of torture: sinners are punished with hammers and other tools.[25] What's more, according to these visions, there is in Hell an even more cruel torture than the directly inflicted physical one: perennial toil. When the devout brother Brendan saw, in the sixth century, a worker on his visit to the beyond, he immediately made the sign of the cross: he realized that he had arrived where all hope must be abandoned. Here is the narrator of his vision:

> When they had passed on further, about a stone's throw, they heard the noise of bellows blowing like

> thunder, and the beating of sledge hammers on the anvils and iron. Then St. Brendan armed himself all over his body with the sign of the Cross, saying, "O Lord Jesus Christ, deliver us from this sinister island." Soon one of the inhabitants appeared to do some work. He was hairy and hideous, blackened with fire and smoke. When he saw the servants of Christ near the island, he withdrew into his forge, crying aloud: "Woe! Woe! Woe!"[26]

If you do not conduct yourself well in this life, the thinking went, you are condemned to work even in the next. And, even worse, that work, according to the pre-Protestant church, will be absolutely useless, meaningless to an extent you could never have imagined even on your worst working day on earth.

This theme crystallizes in the apotheosis of the pre-Protestant worldview, Dante's *Divine Comedy* (completed just before his death in 1321), in which sinners who have devoted their lives to money—both spendthrifts and misers—are doomed to push huge boulders around an eternal circle:

> More shades were here than anywhere above,
> and from both sides, to the sounds of their own screams,
> straining their chests, they rolled enormous weights.
>
> And when they met and clashed against each other
> they turned to push the other way, one side
> screaming, "Why hoard?," the other side, "Why waste?"

> And so they moved back round the gloomy circle,
> returning on both sides to opposite poles
> to scream their shameful tune another time;
>
> again they came to clash and turn and roll
> forever in their semicircle joust.[27]

Dante borrows this idea from Greek mythology. In Tartarus, where the very worst human beings were dispatched, the most severe punishment was meted out to greedy Sisyphus, who was doomed to endlessly push a big rock up to the top of a hill, from which it always rolled back down.[28] Sunday always beckons to Sisyphus and the sinners in Dante's Inferno, but it never comes. They are condemned to an eternal Friday.

Considering this background, we can now gain a better understanding of how great a change in our attitude toward work the Protestant Reformation entailed. In allegorical terms, it moved life's center of gravity from Sunday to Friday. The Protestant ethic reoriented ideology so thoroughly that it even turned Heaven and Hell upside down. When work became an end in itself on earth, the clerics found it difficult to imagine Heaven as a place for mere time-wasting leisure, and work could no longer be seen as infernal punishment. Thus, reformed eighteenth-century cleric Johann Kasper Lavater explained that even in Heaven "we cannot be blessed without having occupations. To have an occupation means to have a calling, an office, a special, particular task to do."[29] Baptist William

Clarke Ulyat put it in a nutshell when he described Heaven at the beginning of the twentieth century: "practically it is a workshop."[30]

The Protestant ethic proved so powerful that its work-centeredness permeated even our imagination. A great example of this is Daniel Defoe's *Robinson Crusoe* (1719), a novel written by a man trained as a Protestant preacher. Marooned on an abundant island, Crusoe does not take it easy; he works all the time. He is such an orthodox Protestant that he does not even take Sunday off, though he otherwise still observes the seven-day week. After saving an aborigine from his enemies, he aptly names him Friday, trains him in the Protestant ethic, and then praises him in a manner that perfectly describes that ethic's ideal worker: "Never man had a more faithful, loving, sincere servant, perfectly obliged and engaged; his very affections were ty'd to me, like those of a child to a father."[31]

In Michel Tournier's twentieth-century satirical retelling of the novel, *Vendredi* (Friday), Friday's conversion to the Protestant ethic is still more total. Crusoe decides to put Friday to the test by giving him a task even more Sisyphean than what Cassian's monastic rule prescribed:

> I set him a task which in every prison in the world is held to be the most degrading of harassments—the task of digging a hole and filling it in with the contents of a second; then digging a third, and so on. He la-

> bored at this for an entire day, under a leaden sky and in heat like that of a furnace. . . . To say that Friday gave no sign of resenting this idiotic employment, is not enough. I have seldom seen him work with such good will.[32]

Sisyphus has truly become a hero.[33]

The Passionate Life

When the hacker ethic is placed in this large historical context, it is easy to see that this ethic—understood not just as the computer hackers' ethic but as a general social challenge—resembles the pre-Protestant ethic to a much greater degree than it does the Protestant one. In this sense, one could say that for hackers the purpose of life is closer to Sunday than to Friday. But, it is important to note, only closer: ultimately, the hacker ethic is not the same as the pre-Protestant work ethic, which envisions a paradise of life without doing anything. Hackers want to realize their passions, and they are ready to accept that the pursuit even of interesting tasks may not always be unmitigated bliss.

For hackers, *passion* describes the general tenor of their activity, though its fulfillment may not be sheer joyful play in all its aspects. Thus, Linus Torvalds has described his work on Linux as a combination of enjoyable hobby and serious work: "Linux has very much been a hobby (but a serious one: the best type)."[34] Passionate and creative, hacking also entails hard work. Raymond says in his guide "How to Become a Hacker": "Being a hacker is lots of fun, but it's a kind of fun that takes a lot of effort."[35] Such effort is needed in the creation of anything even just a little bit greater. If need be, hackers are also ready for the less interesting parts necessary for the creation of the whole. However, the meaningfulness of the whole gives even its more boring aspects worth. Raymond writes: "The hard work and dedication will become a kind of intense play rather than drudgery."[36]

There's a difference between being permanently joyless and having found a passion in life for the realization of which one is also willing to take on less joyful but nonetheless necessary parts.

CHAPTER 2

Time Is Money?

"Time Is Money"

Another central dimension in the hackers' peculiar way of working is their relation to time. Linux, the Net, and the personal computer were not developed in an office between the hours of nine and five. When Torvalds programmed his first versions of Linux, he typically worked late into the night and then woke up in the early afternoon to continue. Sometimes, he shifted from coding Linux to just playing with the computer or to doing something else entirely. This free relation to time has always been typical of hackers, who appreciate an individualistic rhythm of life.

In his famous essay, Weber emphasized the organic connection between the concepts of work and time by incorporating a particular sense of time in his concept of the Protestant work ethic. He quotes Benjamin Franklin's slogan "time is money."[1] The spirit of capitalism arose out of this attitude toward time.

When we think about the network society's dominant relation to time, it is obvious that even though our new economy differs in many other respects from the old industrial capitalism, it largely follows the precepts of the Protestant ethic in regard to the optimization of time. Now, ever shorter units of time are money. Castells aptly speaks about the network society's trend of time compression.[2]

Optimized Time

No one can avoid experiencing the consequences of this optimization of time. The way that our business news is presented is a good cultural indicator of how time pulsates ever more intensely for us. The background music for CNBC's economic news has become more frenetic than that on MTV, and in its speedy visual esthetics it surpasses music videos. Even if one did not understand any of the actual content of the news, one would get the message that there is reason to hurry. Also without understanding the meaning of the news itself, one can realize that it is this speedy economy, the presentation of which on the business-news shows follows the same format as weather reports, that regulates the pace of our actions. In both, we are informed about "weather conditions" to which we simply have to adjust: sunny in New York and a pleasant +80 NASDAQ degrees, typhoon and gain warnings in Tokyo. . . .

In his work *The Information Age,* Castells has demonstrated empirically how competition intensifies in the global information economy (or *informational* economy, to be exact, because all economies are based on information, but ours is based on the new information-technology paradigm; the expression *information economy* will be used as a synonym for this idea).[3] Speedy technological changes make it imperative to get new technology to consumers quickly, before one's competitors do. The slow are left holding obsolete products; even worse is a belated response to fundamental shifts in technology.

Excellent examples of this culture of speed are Amazon.com, Netscape, and Dell Computer, present media symbols of the information economy. Jeff Bezos, a broker who turned into the founder of the Web store Amazon.com, explains the importance of keeping up with technological advance: "When something is growing 2,300 per cent a year [as was the Net at the time of Amazon.com's founding], you have to move fast. A sense of urgency becomes your most valuable asset."[4] Jim Clark, who founded three billion-dollar companies, the second of which was Netscape, describes his flight from Illinois, where the Mosaic browser that was central to the final breakthrough of the Web was created, back to Silicon Valley after he realized the opportunity offered by the Web: "The clock was ticking. Even the three-and-a-half-hour flight from Illinois to San Francisco was lost time. Next to the law of constantly increasing acceleration, Moore's Law, with its eighteen-month increments, seemed almost leisurely [according to Intel founder Gordon Moore, the efficiency of microprocessors doubles every eighteen months].[5] In a lot less time than that, we had to make a whole new product, get it on the market. . . . People weren't going to think in the eighteen-months periods of Moore's Law anymore—that was now an eon!—but in how fast light moved down a fiber optic cable."[6]

Clark's "law of continuous acceleration" compels technological products to be released faster and faster. The capital of successful entrepreneurs in the field also has to move much faster than ever before. Investments frequently change targets within hours, minutes, or even seconds. Capital must not be allowed to stagnate in warehouses or surplus personnel: it must be ready for quick investment in technological innovation or in constantly switching targets in the financial markets.

Time compression has now proceeded to a point where technological and economic competition consists of promising the future to arrive at the consumer faster than it would by the competitor's mediation. New technological inventions are marketed with the claim that they bring us the future now. Correspondingly, in the economic field no one is content to grow wealthy by waiting for the future, which is why the Net companies gain tremendous present worth in record time, long before the realization of their expectations for the future.

In this world of speed, a quick change in the environment (e.g., a technological shift or a surprising fluctuation in the financial market) can cause problems even for excellent enterprises, forcing them to lay off even people who have performed their work superbly.

In order to adjust to these quick changes and accelerated techno-economical competition, enterprises have adopted more agile modes of operation. First, agility is gained by networking. In his Epilogue, Manuel Castells describes the rise of the *network enterprise.*[7] Network enterprises concentrate on their core skills and forge networks according to their changing needs with subcontractors and consultants. It takes too long to acquire every skill oneself, and extra personnel can become a slowing burden later. Network enterprises are even willing to enter project-based alliances with their competitors while continuing otherwise to be energetic rivals. Even internally, network enterprises consist of relatively independent units working together on various projects. People are employed in ways that are more flexible than in the permanent employment model. Castells calls them *flexworkers.*[8] The network model makes it possible for an enterprise to employ only the personnel required for the projects of the moment, which means that in the new economy the real employers are not the enterprises per se but the projects between or within them.[9]

Second, operations in the network society are speeded up by optimization of processes, which is sometimes also called reengineering, after management thinker Michael Hammer's influential *Harvard Business Review* article "Reengineering: Don't Automate, Obliterate" (1990).[10]

Adapting to the new economy does not simply mean adding a webpage to an old process; it involves a rethinking of the entire process. After the change, the process may consist of entirely new stages, but even when that is not the case all unnecessary intermediate stages are dropped and the stopping of products at warehouses is minimized or eliminated. In the culture of speed, immobility is even worse than slowness.[11]

And third, automation, already familiar from industrial society, is still important. It is revealing that the news about high-tech businesses still often shows people at an assembly line. Once a process has been optimized, its parts must still be speeded up by automation (sometimes process-optimization and automation proceed in a reverse order, which can easily lead only to the faster accomplishment of unnecessary or even completely wrong tasks). Even high-tech industry still requires material production, but in it human beings are assigned roles that are as minimal as possible, and they are taught how to perform them in the most time-saving way. So an updated version of Taylorism, the time-optimization method developed by Frederick Winslow Taylor for industrial capitalism, is still alive in the network society.

From the typical information professional of our time, this culture of speed demands an ever more effective use of his or her working hours. The workday is chopped up into a series of fast appointments, and he or she has to hurry from one to the next. Constantly trying to survive some project's deadline, the professional has no time left for playfulness and must optimize his or her time in order to stay on top of it all.

The Fridayization of Sunday

The old Protestant ethic's work-centeredness already meant that there was no time for play in work. This ethic's apotheosis in the information economy can be seen in the fact that the ideal of time optimization is now being extended even to life outside the workplace (if such life still exists). The optimizing pressures on working life—or Friday, if we use the allegories proposed in chapter 1—are now so strong that they begin to eliminate the other pole of the Protestant ethic, the playfulness of free time or Sunday. After the working life has been optimized to its fullest, the requirement of optimality is extended to all of one's other activities, too. Even in time off, one is no longer free merely to "be"—one has to perform one's "being" especially well. For example, only a beginner relaxes without having taken a class in relaxation techniques. To be just a hobbyist in one's hobbies is considered embarrassing.

First, playfulness was removed from work, then playfulness was removed from play, and what is left is optimized leisure time. In his book *Waiting for the Weekend,* Witold Rybczynski provides a good example of the change: "People used to 'play' tennis; now they 'work' on their backhand."[12] Another work-centered way to spend leisure is to engage in the practice of skills important to work or else to detach oneself as optimally as possible from work in order to be able to continue it in the best possible shape.

In an optimized life, leisure time assumes the patterns of work time. Home time is scheduled and planned as tightly as work time: Take child to sports practice 5:30–5:45. Gym 5:45–6:30. Therapy session 6:30–7:20. Pick up child from practice 7:20–7:35. Prepare food and eat 7:35–8:00. Watch television with family 8:00–11:00. Put child to bed. Converse with spouse 11:00–11:35. Watch late-night show 11:35–12:35. Other attention paid to spouse (occasionally) 12:35–12:45. The day is divided on a business model in clear-cut segments of time, and that division is of course reinforced by television program schedules. The time spent at home is often experienced in a way similar to the way time at work is experienced: rushing from one appointment to another so that one manages to keep them all. Aptly, one mother explained to an interviewer how she felt that families now have a new status symbol: "It used to be a house or a car. Now you say, '*You're* busy? You should see how busy *we* are.' "[13]

In *Time Bind,* sociologist Arlie Russell Hochschild gives an excellent description of the extent to which the home has started using business methods to optimize time. Hochschild does not examine these changes in the home in relation to the information economy, but it is easy to place these changes in their larger context by regarding them as adaptations of the three forms of time optimization used in business life. The home, too, has been Taylorized or automated to make the human being's task as simple and as quickly performable as possible. Hochschild speaks aptly of "deskilling parents at home": microwavable prepared foods have replaced homemade dinners based on personal recipes. Families no longer create their own entertainment but simply punch the remote to tune themselves into television's social assembly line. Hochschild's irony is accurate: "After dinner, some families would sit together, mute but cozy, watching sitcoms in which television mothers, fathers, and children talked energetically to one another."[14]

In home-life management, another business strategy comes into play: networking, especially in the form of outsourcing, from take-out food to day-care centers (subcontracting food production and child care). Hochschild gives a good description of the resultant new image of the mother (or father): "The time-starved mother is being forced more and more to choose between being a parent and buying a commodified version of parenthood from someone else. By relying on an expanding menu of goods and services, she increasingly becomes a manager of parenthood, supervising and coordinating the outsourced pieces of familial life."[15]

Third comes the optimization of process. Even at home, the "process" of child care is optimized by eliminating its "unnecessary" parts. No longer do parents just hang out inefficiently with the children; they spend "quality time" with them. This quality time is clearly defined in terms of its beginning and end, and in the course of it some event clearly takes place or some concrete outcome is achieved (e.g., the child's school play or athletic contest, or a trip to the amusement park). In quality time, all downtime is minimized or obliterated. A parent who has completely internalized the culture of speed may even believe that the child, too, experiences this as equal to or even better than a relationship in which the adult has unconditioned time for the child. Hochschild comments: "Quality time holds out the hope that scheduling intense periods of togetherness can compensate for an overall loss of time in such a way that a relationship will suffer no loss of quality."[16]

Flexible Time

In the information economy, all of life has become more optimized in a way typical (and in former times not even typical) of work. But this is not all. In addition to the work-centered *optimization* of time, the Protestant ethic also means the work-centered *organization* of time. The Protestant ethic introduced the idea of regular working time as the center of life. Self-organization was lost and relegated to a region of work's leftovers: the evening as what's left of the day, the weekend as the remainder of the week, and retirement—the leftovers of life. At the center of life is the regularly repeated work, which organizes all other uses of time. Weber describes how, in the Protestant ethic, "irregular work, which the ordinary labourer is often forced to accept, is often unavoidable, but always an unwelcome state of transition. A man without a calling thus lacks the systematic, methodical character which is . . . demanded by worldly ascetism."[17]

So far, this organization of time has not changed a lot in the information economy. Few, still, can deviate from strictly regular working hours, despite the fact that the new information technologies not only compress time but also make it more flexible. (Castells calls this the "desequencing of time.") With technologies like the Net and the mobile phone, one can work where and when one wants.

But this new flexibility does not automatically lead to a more holistic organization of time. In fact, the dominant development in the information economy seems to be that flexibility is leading to the strengthening of work-centeredness. More often than not, the information professionals use flexibility to make leisure time more available for brief spells of work than the other way around. In practice, the block of time reserved for work is still centered on an (at least) eight-hour workday, but leisure time is interrupted by spells of work: half an hour of television, half an hour of e-mail, half an hour outside with the kids, interspersed with a couple of work-related mobile-phone calls.

Wireless technology—such as the mobile phone—is

not in itself a technology of freedom; it can be an "emergency technology" as well. It easily happens that every call turns into an urgent call, and the mobile phone becomes a tool for surviving the day's emergencies.

Against this background, there is appropriate irony in the fact that the first adopters of the phone (both landline and radio) were emergency professionals, such as policemen who needed to respond to urgent situations. Aronson and Greenbaum describe how, for example, wired doctors "gradually but steadily assumed the moral obligation to be reachable by telephone at all times."[18] Even to the larger public, the phone was originally marketed as a survival tool. A 1905 advertisement described how the phone can save a lonely housewife's life: "The modern woman finds emergencies robbed of their terror by the telephone. She knows she can summon her physician, or if need be, call the police or fire department in less time than it ordinarily takes to ring for a servant."[19] Another marketing point was that a businessman could call his wife to tell her that he is late because of some urgent matter. In a 1910 advertisement a man says to his wife, "I'll be half an hour late," and the wife replies cheerfully, "All right, John." The text below the picture explains further: "Unexpected happenings often detain the business man at his office. With a Bell telephone on his desk and one in his home, he can reach his family in a moment. A few words relieve all anxiety."[20]

Since the first words on the phone by its inventor,

Alexander Graham Bell, to his assistant in 1876 ("Mr. Watson, come here, I want you"), the phone has been linked to a culture of urgency. The paradox is that the highest technology brings us easily to the lowest level of survival life, in which we are constantly on call, reacting to urgent situations. There is a strong tendency in this direction in the information-economy elite's image: in the past, you belonged to the elite when you no longer had to run from one place to the next, working all the time; nowadays, the elite consists of people perennially on the move, taking care of urgent business on their mobile phones and always trying to survive some deadline.

The Sundayization of Friday

If we use the new technology to further work-centeredness, technologies such as the mobile phone easily lead to a work-centered dissolution of the boundary between work and leisure. *Both* the optimization and flexibility of time may lead to Sunday becoming more and more like Friday.

But this is not inevitable. Hackers optimize time to be able to have more space for playfulness: Torvalds's way of thinking is that, in the middle of the serious work of Linux development, there always has to be time for some pool and for some programming experiments that do not have immediate goals. The same attitude has been shared by hackers since the MIT of the sixties. In the hacker version of flextime, different areas of life, such as work, family, friends, hobbies, et cetera, are combined less rigidly, so that work is not always at the center of the map. A hacker may join his friends in the middle of the day for a long lunch or go out with them for a beer in the evening, then resume work late in the afternoon or the next day. Sometimes he or she may spontaneously decide to take the whole day off to do something completely different. The hacker view is that the use of machines for the optimization and flexibility of time should lead to a life for human beings that is less machinelike—less optimized and routine. Raymond writes: "To behave like a hacker, you have to believe this [that people should never have to drudge at stupid, repetitive work] enough to want to automate away the boring bits as much as possible, not just for yourself but for everybody else." When the hacker ideal of more self-determined use of time becomes realized, Friday (the workweek) should become more like what Sunday (the "leftovers of life") has traditionally been.

Historically, this freedom to self-organize time again has a precursor in the academy. The academy has always defended a person's freedom to organize time oneself. Plato defined the academic relation to time by saying that a free person has *skhole,* that is "plenty of time. When he talks, he talks in peace and quiet, and his time is his own."[21] But *skhole* did not mean just "having time" but also a certain relation to time: a person living an academic life could *organize one's time oneself*—the person could combine work and leisure in the way that he wanted. Even though a free person could commit to doing certain works, no one else owned his time. Not having this charge of one's time—*askholia*—was associated with the state of imprisonment (slavery).

In the pre-Protestant life, even outside of the academy, people were more in charge of their time than after the Protestant Reformation. In his book *Montaillou: Cathars and Catholics in a French Village, 1294–1324,* Emmanuel Le Roy Ladurie creates a fascinating portrait of life in a medieval village at the turn of the thirteenth and fourteenth centuries. The villagers had no way to define time in any exact way. When they spoke of it, they used vague expressions, saying that something had happened "at the season when elms have put forth their leaves" or that something took "the time it takes to say two Paternosters."[22] In Montaillou, there was no need for more exact time measurements, since the village was not run according to any regular working time.

Le Roy Ladurie writes: "The people of Montaillou were not afraid of hard work and could make an effort if necessary. But they did not think in terms of a fixed and continuous timetable. . . . For them the working day was punctuated with long, irregular pauses, during which one would chat with a friend, perhaps at the same time enjoying a glass of wine. *At those words,* said Arnaud Sicre, *I folded up my work and went to Guillemette Maury's house.* And Arnaud Sicre indicates several other similar interruptions: *Pierre Maury sent for me in the shop where I made shoes. . . . Guillemette sent a message to ask me to go to her house, which I did. . . . Hearing that, I left what I was doing.*"[23]

In Montaillou, it was still, to a great extent, the worker and not the clock that determined the pace. Nowadays, a shoemaker who decided to go off and have a glass of wine with a friend in the middle of the day would be fired no matter how many shoes he produces and how excellently. This is because the workers of our time no longer enjoy the same freedom to manage their own time a cobbler or shepherd enjoyed in the "dark" Middle Ages. Of course, no picture of medieval work is complete without mentioning that there was also land slavery, but if we allow this important exception, we can say about medieval work that as long as reasonable goals were met, no one supervised the workers' use of time.

Only in monasteries was activity tied to the *clock,* so, once again, the Protestant ethic's historical precedent can be found in the monastery. In fact, when reading the monastic rules, one often feels that one is reading a description of the dominant contemporary company practices. Benedict's rule is a good example. It taught that life's pattern must "always be repeated at the same Hours in just the same way."[24] These "Hours" were the seven canonical Office Hours (*horas officiis*):[25]

dawn	Lauds (*laudes*)
9 A.M.	Prime (*prima*)
noon	Sext (*sexta*)

3 P.M.	None (*nona*)
6 P.M.	Vespers (*vespera*)
nightfall	Compline (*completorium*, the completion of the day)
at night	Matins (*matutinae*)

The canonical hours circumscribed the time for all activities. Following them, wake-up time was always the same, as was bedtime.[26] Work, study, and meals were also assigned exact hours.

Under Benedict's rule, deviation from the schedule set for life was a punishable act. Oversleeping was condemned: "Let every precaution be taken that this does not occur."[27] No one was allowed to take a break for a self-decided snack: "Let no one presume to take food or drink before or after the appointed time."[28] Missing the beginning of the sacred Office Hours was punished[29]—the only exception to the demand for absolute promptness in regard to Office Hours was the night prayer, to which one could arrive at any time until the reading of the second psalm (a "staggered work hour").[30]

The Protestant ethic brought the clock out of the monastery into everyday life, giving birth to the concept of the modern worker and the notions of workplace and work time associated with it. After that, Franklin's words in his autobiography applied to all: "Every part of my business should have its allotted time."[31] Despite its new technology, the information economy is still predominantly based on Office Hours, with no place for individual variations.

This is a strange world, and the shift to it did not take place without strong resistance. In his article "Time, Work-Discipline, and Industrial Capitalism" (1967),[32] social historian Edward Thompson characterizes the difficulties encountered in the transition to industrial work. He notes that the medieval agriculturists, for example, were used to task-oriented work. In their traditional thinking the essential thing was to complete their tasks. The weather set exterior limits, but, within these, tasks could be dealt with according to individual inclination. Industrial work, on the other hand, was time-oriented: work was defined by the time used for it. It was this idea of defining a work relation with time and not the works per se that preindustrial people found alien, and they resisted it.

The interesting promise of the new information technology is that it could make a new form of task-oriented work possible. But it is important to remember that this does not happen automatically. In fact, the strange truth is that at the moment this technology is used more for the intensified supervision of a worker's time, through devices such as the time clock. (The absurdity of this application of technology brings to my mind the educative month I spent in industrializing India. On my daily walks, I began to pay attention to Indian street sweepers who were on the corners from morning to night, but the streets never seemed to get any cleaner. When I expressed my puzzlement to an Indian friend and asked why the managers of these sweepers do not complain about the situation, he answered that I had looked at the matter from an entirely false perspective. I had assumed, erroneously, that the Indian street sweeper's task was to sweep the streets, but—he added—the Indian street sweeper's job is not to sweep the street; it is to *exist impeccably in the capacity of a street sweeper*! This is a good expression for the ideology that is also behind the time clock. The most refined time clocks that I have seen have dozens of codes that people are expected to use in order to indicate with which particular nuance of their impeccable existing they are engaged at any given time, including the state of their digestive system, which is the main justifier of pauses. This is time-oriented use of technology at its purest.)

The Rhythm of Creativity

One cannot deny that our management still focuses too much on the external factors of work, like the worker's time and place, instead of inciting the creativity on which the company's success depends in the information economy. Most managers have not understood the deep consequences of the question, Is our purpose at work to "do time" or to do something? In the early seventies, Les Earnest of the artificial-intelligence laboratory at Stanford University gave a good précis of the hackers' answer to this question: "We try to judge people not on how much time they waste but on what they accomplish over fairly long periods of time, like a half year to a year."[33]

This answer can be understood both purely pragmatically or ethically. The pragmatic message is that the information economy's most important source of productivity is creativity, and it is not possible to create interesting things in a constant hurry or in a regulated manner from nine to five. So even for purely economic reasons, it is important to allow for playfulness and individual styles of creativity since, in the information economy, the culture of supervision turns easily against its desired objectives. Of course, an important added condition is that in the realization of the task-oriented project culture, set project schedules are not too short-term—that they are not the *dead*lines of the survival life—so that there is a genuine opportunity for creative rhythm.

But, of course, the ethical dimension involved here is even more important than these pragmatic considerations: we are talking about a worthy life. The culture of work-time supervision is a culture that regards grown-up persons as too immature to be in charge of their lives. It assumes that there are only a few people in any given enterprise or government agency who are sufficiently mature to take responsibility for themselves and that the majority of adults are unable to do so without continuous guidance provided by the small authority group. In such a culture, most human beings find themselves condemned to obedience.

Hackers have always respected the individual. They have always been anti-authoritarian. Raymond defines the hacker position: "The authoritarian attitude has to be fought wherever you find it, lest it smother you and other hackers."[34]

The hacker ethic also reminds us, in the midst of all the curtailment of individual worth and freedom that goes on in the name of "work," that our life is here and now. Work is a part of our continuously ongoing life, in which there must be room for other passions, too. Reforming the forms of work is a matter not only of respecting the workers but of respecting human beings as human beings. Hackers do not subscribe to the adage "time is money" but rather to the adage "it's my life." And certainly this is now our life, which we must live fully, not a stripped beta version of it.

PART TWO

The MONEY *Ethic*

CHAPTER 3

Money as a Motive

The Money Ethic

As we have seen, the hacker ethic means a *work ethic* that challenges the prevailing Protestant ethic. It may not be that difficult to agree with much of the hackers' challenge for work—in fact, despite the fact that the Protestant work ethic still has a strong hold over the information economy, the hacker work ethic seems to be slowly spreading from computer hackers to the larger group of information professionals. But when we come to the second main level of Weber's concept of the Protestant ethic—the *money ethic,* our relation to money—reactions are bound to be more divided.

Writing about this dimension of the spirit of the old capitalism, the Protestant money ethic, Weber said: "The *summum bonum* of this ethic," its highest good, is "the earning of more and more money."[1] In the Protestant ethic, both work and money are seen as ends in themselves.

The "new economy's" "newness" does not consist of rejecting the old goal of moneymaking. Truth be told, we are living in the most purely capitalist era of history, of which it is an apt little symbol that the traditional counterbalance of the capitalist spirit, the antimarket-spirited Sunday, feels so alien to us that we want to get rid of the remaining Sunday shop closings and turn Sunday into another Friday. The change in our relation to Sunday is also a good sign of a related important shift in the Protestant ethic in the new economy: Sunday, meaning leisure, has a place mainly as a space for consumption. Weber's frugal seventeenth-century Puritan has been replaced by the omnivorous twenty-first-century gratification-driven consumer.

This means that the Protestant ethic's central conflict is now resolved in a new way. The conflict arose from the simultaneity of the demand for work that furthers economic prosperity *and* the demand to regard any kind of work at all as a duty. But if a person really sees work as the highest value, he or she does not worry about the maximization of his or her income. And if a person regards money as the highest goal, work is no longer a value in itself but merely a means. In the old capitalism, this conflict was resolved by ranking work higher than money, which is reflected in the way most people tend to understand the term *Protestant ethic* as *Protestant work ethic.*

In the new economy, work is still an autonomous value,

but it is subordinated to money. Of course, there are still many people who consider work to be a higher value, and societies still tend to condemn the idle, even when they are wealthy enough not to need to work. But gradually the balance between work and money is tipping in favor of the latter, enticed by the way in which wealth is accumulated in the new economy. The financial result of work produced by a business (its dividends) is becoming less important than its capital growth, the increase in its stock's value. The relationship between work (pay) and capital is shifting in favor of capital. This is a result of stock options, start-up business, shares as a form of reward, and individuals putting less money in the bank in favor of investing in the stock market. Where the seventeenth century's work-centered Protestants specifically banned betting, the new economy depends on it.

In addition to strengthening the position of money, the new economy similarly strengthens the idea of *ownership*, which is central to the old spirit of capitalism, by extending it to information to an unprecedented degree. In the information economy, companies realize their moneymaking goal by trying to own information via patents, trademarks, copyrights, nondisclosure agreements, and other means. In fact, information is guarded to such an extent that when one visits an information-technology company, sometimes one cannot avoid the impression that all these locks protecting information make the building similar to a maximum-security prison.

In stark contrast to this revitalized Protestant money

ethic, the original computer-hacker ethic emphasized openness. As was mentioned, according to the hackers' "jargon file" the hacker ethic includes the belief that "information-sharing is a powerful positive good, and that it is an ethical duty of hackers to share their expertise by writing free software."[2] Whereas the historical precursor for controlling the free flow of information is the monastery (in his rule, Benedict elevated to the status of principle a quote from the Bible that would work well for many new-economy enterprises: "Keep silence even from good things,"[3] and in monasteries the drive for freedom of information, *curiositas,* was regarded as a vice),[4] the historical precedent of the hacker ethic is the academic or scientific ethic (when sociologist of science Robert Merton gave his famous expression to the scientific ethic's development in the Renaissance, he emphasized that one of its cornerstones was "communism," or the idea that scientific knowledge must be public[5]—(an idea that the Renaissance revived from the academic ethic of the first scientific community, Plato's Academy, which was based on the idea of *synusia,* concerted action in which knowledge was shared freely).[6]

In line with this hacker ethic, many hackers still distribute the results of their creativity openly, for others to use, test, and develop further. This is true about the Net, and Linux is another good example. It has been created by a group of hackers who have used their leisure time to work together on it. To ensure the preservation of its open development, Torvalds "copylefted" Linux from the beginning. ("Copyleft" is a form of licensing originally developed in Stallman's GNU project, which guarantees that all developments will be available for free use and further development by others. Stallman picked the name from a line on the envelope of a letter he received: "Copyleft: all rights reversed.")[7]

Money as a Motive

In the midst of a time in which the money motive has become stronger and led to the closing off of more and more information, it is surprising to see how these hackers explain why they undertake so huge a project as Linux in which money is not a driving force, but, instead, in which creations are given away to others. At the beginning of this book, Torvalds presents his "Linus's Law" for positioning this form of hackerism in the context of general human motives. Conscious of the simplification, he talks about three *ultimate* motives, which he calls *survival, social life,* and *entertainment. Survival* is mentioned only briefly as the lowest level, as a prerequisite to fulfilling the higher motives. In this book's vocabulary, Torvalds's *entertainment* corresponds to *passion:* it is the state of being motivated by something intrinsically interesting, enticing, and joyful.

Social life encompasses the need for belonging, recognition, and love. It is easy to agree that these are fundamental forces. Every one of us needs to belong to some group within which we feel approved. But mere approval is not enough: we also want to be recognized for what we do, and we have a need for an even deeper experience, that of feeling loved and loving someone else. To put it in another way, the human being needs the experience of being part of a *We* with some others, the experience of being a respected *He* or *She* within some community, and the experience of being a special *I* with someone else.

Many hackers have expressed similar views since the sixties. Wozniak, for example, summed up the elements that motivate his action in the speech he gave upon graduating from the University of California at Berkeley in 1986: "You don't do anything in life unless it's for happiness. . . . That's my theorem of life. . . . A simple formula, really: $H = F^3$. Happiness equals food, fun, and friends."[8] (In Wozniak's terminology, *food* corresponds to Torvalds's *survival*, *friends* to *social life*, and *fun* to *entertainment*.) And, of course, this hacker view resembles very much some attempts in psychology to classify the most fundamental human motivations—especially the five-level hierarchy of needs described in Abraham Maslow's *Motivation and Personality* (1954) and *Toward a Psychology of Being* (1962). This hierarchy is often represented as a pyramid, the top of which represents our highest motives. On the bottom level, we find physiological needs, the need to survive, which is closely connected to the second level, the need to feel safe. The third level calls for social belonging and love, and it is closely connected to the fourth level, the need for social recognition. The highest level calls for self-realization. It is not hard to see how Torvalds's triad of survival, social life, and entertainment corresponds to Maslow's model.

Inevitably, such simplifications ignore the psychological multifariousness of human action, but, given that theoretical caveat, Torvalds's/Maslow's model can nevertheless cast some light on how these hackers' motivation for action differs from motivation in the Protestant ethic. "Survival" or "You have to do something to earn your living" is the answer a great number of people will give when asked why they work (often responding in a mildly puzzled fashion, as if this went without saying). But strictly speaking, they do not mean mere survival—that is, having food and so on. In their use, *survival* refers to a certain socially determined lifestyle: they are not working merely to survive but to be able to satisfy the form of social needs characteristic to a society.

In our society infused by the Protestant ethic, work is actually a source of social acceptance. An extreme example of this can be found in the philosopher Henri Saint-Simon's nineteenth-century Protestant plan of an ideal society: only those who work are counted as citizens—a complete contrast to the ideal societies of antiquity, such as the one presented by Aristotle in his *Politics,* in which only those who did not have to work were considered worthy of citizenship.[9] Even when the work itself does not involve social interaction, social acceptance beyond mere breadwinning remains an important social motivation for working.

Of course, in almost every kind of work, the need to belong also finds its expression within the particular social circumstances of the workplace, as people have opportunities to participate in social exchanges with both fellow workers and clients. In the workplace, people can gossip, discuss their living situations, and argue about current events. By doing a good job, a person may also gain recognition. And the workplace is even a forum for falling in love. Naturally, these social motives as such were also intertwined with work before the Protestant ethic, but this ethic did entail a new, peculiar way of realizing them. In a work-centered life governed by the Protestant ethic, people hardly have friends outside of their work, and there are few other places for falling in love. (Think of the number of people who now find a spouse among their colleagues or other people who met in work-related circumstances, and how frequent workplace romances are.) In this lifestyle, life outside of work often does not provide the social belonging, recognition, or love traditionally experienced in the home or at leisure, and therefore work easily turns into a surrogate for home—which does not mean that work now takes place in a relaxed "home" atmosphere but that a person needs work to satisfy these motives, because work-centeredness has invaded and annexed leisure life.

In the hacker community, social motivations play an important part but in a very different way. One cannot actually understand why some hackers use their leisure for developing programs that they give openly to others without seeing that they have strong social motives. Raymond says that these hackers are motivated by the force of *peer recognition.*[10] For these hackers, recognition within a community that shares their passion is more important and more deeply satisfying than money, just as it is for scholars in academe. The decisive difference from the Protestant ethic is that for hackers it is important that peer recognition is no substitute for passion—it must come as a *result* of passionate action, of the creation of something socially valuable to this creative community. Under the Protestant ethic, the opposite is often the case: social motivations serve to distract attention from the idea that work itself should involve the realization of a passion. As a result, the Protestant ethic's emphasis on the social features of work becomes a double surrogate: for the lack of social life outside of work *and* for the absence of an element of passion in the work itself.

It is this hackers' linking of the social level to the passionate level that makes their model so powerful. Hackers realize something very important about the most deeply satisfying social motives and their potential. In this, hackers contradict the stereotypical image of hackers' asociability—a stereotype that has never been very true. (Marvin Minsky, the famous AI researcher at whose lab the first MIT hackers programmed, perhaps thinking about this same phenomenon, even said of them, "Contrary to common belief, hackers are more social than other people.")[11]

The Protestant ethic's pursuit of work and money is also based on these same three categories of social motives, but because in it the satisfaction of social needs is mediated by money and work and does not derive directly from the nature of activity and its creations, it cannot bring about the same effect. The consequence is that when social motives do not find an ally in passion, they become allied with survival, and life becomes concentrated on "making a living."

Hackers like Torvalds, proponents of passion and community, find such a life, permeated with the lowest survival tone, very strange. There is, indeed, reason to wonder why, in spite of all our technological advances, people's days are still so predominantly devoted to what they call breadwinning. Shouldn't this incredible technological evolution have raised us from the survival level to higher ones? Perhaps we should see the dominant progress as the history not of making our lives easier but of making breadwinning continuously more difficult. As the Chinese philosopher Lin Yutang has commented, from the perspective of the civilization governed by the Protestant ethic, "Civilization is largely a matter of seeking food, while progress is that development which makes food more and more difficult to get."[12]

There is a great difference between choosing a field of study or answering a want ad on the basis of maximizing income *and* first considering what one really would like to do with one's life and only then pondering how to make that financially possible. For hackers like Torvalds, the basic organizational factor in life is not work or money but passion and the desire to create something socially valuable together.

This primary question of life organization is immensely important. If making money is the main goal, a person can often forget what his or her true interests are or how he or she wants to deserve recognition from others. It is much more difficult to add on other values to a life that started out with just making money in mind than it is to make some personally interesting endeavor financially possible or even profitable. In the first case, the thing I am doing even though I find it uninteresting is in all likelihood equally uninteresting to others, and in order to sell it to them I have to persuade them to believe that this intrinsically uninteresting thing is interesting after all (the task of most advertising).

Capitalist Hackers

That said, one should not think of most hackers' attitude toward money as either some paradisiacal utopianism or some kind of essential aversion to it. The original hacker ethic was primarily a matter of what place money is accorded as a motive and what types of its influence on other motives should be avoided. Hackers are not naïve. They are not blind to the fact that in a capitalist society it is actually very difficult to be completely free unless a person has sufficient individual capital. The capitalist gains power over the lives of others by means of money. It is exactly when working for someone else that a person may not be free to base her or his work on personal passion, a person loses the right to determine one's life rhythms, and the ideal of openness is not in one's own power. But if one is the empowered capitalist, one can make one's own life decisions.

There are many examples of hackers who have chosen "capitalist hackerism." Some take part in the traditional capitalism only temporarily: these hackers generate financial independence by shares or stock options acquired through running a company or by working for some years around his or her passion. Wozniak is a good example. When, at the age of twenty-nine, Woz retired from Apple only six years after its founding, he owned shares valued at some one hundred million dollars (even after having sold a considerable number of his shares at an extraordinarily low price to fellow workers because he wanted to spread the wealth more fairly within the enterprise).[13] Thanks to his financial independence, Wozniak has since been able to freely choose his actions. He describes his life after Apple as follows: "I've got accountants and secretaries to handle everything so I can spend as much time as I can doing what I like to do, which is to work with computers and schools and kids."[14] After leaving Apple, Wozniak decided to go back to college to fulfill the formal requirements needed for realizing his dream of teaching new generations of hackers. (He teaches the use of computers to children in the local schools and at his home.)

There are also hackers who think that being a hacker consists primarily of passionate action and the freedom to organize one's time and that as long as this work ethic is realized there is no problem with making money permanently through traditional capitalism. Many of the best-known technological businesses serve as good examples. The group of young people that founded Sun Microsystems in 1982 to design networked work-stations consisted of Berkeley's Bill Joy and three Stanford students, including the German-born technological wiz Andreas "Andy" Bechtolsheim. The name of their business was an acronym for Stanford University Network, on which Bechtolsheim had been working. Bechtolsheim reminisces about the passion shared by the original crew: "We were twenty-something-year-olds running a company and we had just met, but we certainly shared the passion."[15] Both Joy and Bechtolsheim have continued in the business environment: Joy has stayed on to navigate Sun, and Bechtolsheim has moved to another hacker-created enterprise, the Internet router manufacturer Cisco Systems. It is through these kinds of technological enterprises started by hackers that the hacker work ethic is slowly spreading to other kinds of businesses, just as, according to Weber, the Protestant ethic in its time grew from influencing enterprises started by Protestants to becoming the ruling spirit of capitalism.

But there is an inherent tension in the idea of hackerism within a very traditional capitalism. The original meanings of the terms *capitalism* and *hacker* pull in different directions. In tune with the Protestant ethic's focus on money, the supreme goal of capitalism is the increase of capital. The hackers' work ethic, on the other hand, emphasizes passionate and free-rhythmed activity. Even though it is theoretically possible to reconcile both goals, the tension between them is often resolved in practice by dropping the hackerism and just following the guidelines of the Protestant ethic.

The computer hacker's number-one enemy, Bill Gates's Microsoft, serves as a good example. When Gates cofounded the company in 1975, he was just a hacker like Joy, Wozniak, or Torvalds. Computers had been his passion from childhood, and he had used all the time available to him programming on the local Computer Center Corporation's computer. Gates even gained hacker respect by programming his first interpreter of the BASIC programming language without access to the computer for which it was intended (the MITS Altair): it worked. With his friend Paul Allen, Gates founded Microsoft with the express initial intention of creating programming languages for personal computers, which was a very hackerist starting point, since only hackers used these machines for programming.[16]

In Microsoft's subsequent history, the profit motive has taken precedence over the passion. Since capitalist hackerism shares the Protestant ethic's goal of maximizing money, this focus is bound to influence and finally dominate the work ethic of an enterprise. When money becomes the highest end in itself, passion is no longer an essential criterion for work choices. Projects are chosen primarily on the basis of the greatest promise of profit. Recognition, then, is determined by one's power position—one's place within the organization and one's personal wealth.

After Microsoft's start-up phase, Gates has occasionally described his attitude toward work in tones that sound much more like the Protestant than the hacker ethic. For example: "If you don't like to work hard and be intense and do your best, this is not the place to work."[17]

Free Market Economy

Given the problems of combining hackerism and the current form of capitalism, a group of hackers is going in new directions to defend a new type of economy, based on the so-called open-source enterprise that develops software on the open model. In this model, exemplified by such successful companies as the Linux developer Red Hat, anyone is free to learn by studying the source code of these programs and even to develop them further into one's own open products.[18] The spiritual father of these companies is the controversial Richard Stallman, whose thinking is so radical that many of the actual open-source companies prefer to keep their distance from him as a person. A typical expression of Stallman's uncompromising approach is his "Free Software Song," which he has recorded for the Net:

> Join us now and share the software;
> You'll be free, hackers, you'll be free.
> (*repeat*)
>
> Hoarders may get piles of money,
> That is true, hackers, that is true.
> But they cannot help their neighbors;
> That's not good, hackers, that's not good.
>
> When we have enough free software
> At our call, hackers, at our call,
> We'll throw out those dirty licenses
> Ever more, hackers, ever more.
>
> Join us now and share the software;
> You'll be free, hackers, you'll be free.
> (*repeat*)[19]

To many, this may initially sound like a form of communism or even utopianism. But a closer look reveals that it is actually neither. Despite its apparent anticapitalist tone, Stallman's hackerism does not actually oppose capitalism as such. Stallman says that the word *free* as he uses it in *free software* in his song and other more serious writings does not necessarily mean "free of charge" but simply "freedom." He suggests interpreting the idea in the sense of *free speech,* not *free beer.*[20] Stallman's version of the hacker money ethic does not oppose *making money,* just making money *by closing off information from others.* He is proposing a new kind of free-market economy: a *free* market economy in a much deeper sense than in the normal capitalist vocabulary, but still a capitalist economy. It is this radical idea that is the hardest for many open-source companies to follow, and they prefer basing their open model on purely pragmatic argument: the open-source model is chosen for those projects in which it is superior in technical or economic terms; otherwise, the closed model is preferred.[21]

In Stallman's ethical approach, the stakes are higher. Its question is, Is the present company practice of restricting information really ethically tenable? The fact that it is the current model does not make it the right one or mean that it has been argued for soundly. One rarely hears anyone trying to make an intellectually satisfying case for the present practice without any changes. Any serious attempt should address many fundamental issues of our information age, including, for example, the paradoxical dependence of closed information on open information. This paradox is at the heart of our time: in fact, if one takes technology companies' dependence on research seriously, one might say that the ethical dilemma facing businesses in the new information economy is that capitalist success is possible only as long as most of the researchers remain "communists" (in Merton's sense). Only as long as scientific knowledge is left open do marginal secret additions to the collective information lead to dramatic individual gains. This paradox is due to the fact that the network society is not determined only by capitalism but to an at-least-equal degree by scientific "communism." A Stallman-esque hacker might be inspired to proclaim, "Present capitalism is based on the exploitation of scientific communism!" Receiving the information produced by everyone else while withholding all the information produced by oneself presents an ethical quandary. This quandary grows worse with the progress of the information age, since an ever greater part of products' value derives from their underlying research.

The question that this extreme form of the hacker ethic teases us with is this: Could there be a free market economy in which competition would not be based on controlling information but on other factors—an economy in which competition would be on a different level (and, of course, not just in software but in other fields, too)? In answering this question, we should not try to get around it with an easy, and faulty, solution by saying that this is a new form of communism, which we have seen does not work. It is not really *communism:* communism involves a centralized authority model—communism is a form of statist economy—and that is alien to hackers. (Thus, Merton's choice of the label *communism* for one main characteristic of the scientific ethic is an unfortunate one because he means by it a totally different idea: the openness of information.)

In addition, when the hacker work ethic opposes capitalism's work-centeredness, it also opposes the same feature in communism. One must remember that despite their major differences, both capitalism and communism are based historically on the Protestant ethic, as sociologist Peter Anthony has reminded us in *The Ideology of Work:* "All these ingredients identified in the Protestant ethic [behind capitalism]: work, measurement, rationalism, materialism, are present [in communism] not as confused alternatives to other and more widely accepted notions, but as dominant themes which demand that others must be removed."[22] Seen from this angle, the CEO in his rolled-up shirtsleeves does not differ greatly from the Soviet labor hero wielding his sickle in the fields: both of them are champions of work. Capitalism, communism, and the new information economy so far each merely propagate the form of the Protestant ethic that each considers the purest.

All forms of the hacker money ethic mean a challenge

to all the existing systems. The hacker community is not unified in its answers to these big questions, but even just having started a debate on these questions in the nucleus of the information economy is a radical enough challenge.

CHAPTER 4

The Academy and the Monastery

The Open Model

In the original hacker money ethic, the new economy's governing attitude, "which seeks profit rationally and systematically" (Weber's description of the spirit of old capitalism, which still applies well to our time),[1] is challenged by the open model in which the hacker gives his or her creation freely for others to use, test, and develop further. For the original MIT hackers, this idea was even as defining an element of the hacker ethic as the hacker relation to work, but nowadays the "jargon file" says that this ethical ideal of openness is accepted among hackers "widely, but not universally."[2]

Although from this book's perspective the ethical arguments of hackerism are the most interesting and important ones, there is also a more pragmatic level that is significant and fascinating. Just as we can add to our ethical arguments for the passionate and free work ethic the more pragmatic point that, in the information age, new information is created most effectively by allowing for playfulness and for the possibility of working according to one's individual rhythm, we can likewise say that the open model is not just ethically justified but also very powerful in practice. (In fact, the "jargon file" also says that it is a "*powerful* positive good.") It is worth taking a closer look at the hackers' idea of openness from this viewpoint. The development of the Net would be a great example, but the Linux project, which has arguably taken the ideal of openness the furthest so far, serves as an even better one. After understanding this powerful model that has made the Net and Linux possible, we can think of some ways in which the open model could be applied to areas of life other than software.

Torvalds started working on Linux in 1991 while he was a student at the University of Helsinki.[3] After developing an interest in the problems of operating systems, Torvalds imported into his home computer the Unix-like Minix operating system, written by Dutch computer-science professor Andrew Tanenbaum and, by studying and using it as a developmental framework, proceeded to design his own one.[4] An essential feature of Torvalds's work was that he involved others in his project from the very beginning. On August 25, 1991, he posted a message on the Net with the subject line "What would you like to see most in minix?" in which he announced that he was "doing a (free) operating system."[5] He received several ideas in reply and even some promises for help in testing the program. The operating system's first version was released on the Net as source code free to all in September 1991.[6]

The next, improved version was available as soon as early October. Torvalds then extended an even more direct invitation to others to join him in the development of the new system.[7] In a message sent to the Net, he asked for tips about information sources. He got them, and development advanced quickly. Within a month, other programmers had joined in. Since then, the Linux network has grown at an amazing creative pace. Thousands of programmers have participated in Linux's development, and their numbers are growing steadily. There are millions of users, and their number, too, is growing. Anyone can participate in its development, and anyone is welcome to use it freely.[8]

For the coordination of their development work, Linux hackers use the entire toolbox of the Net: e-mail, mailing lists, newsgroups, file servers, and webpages.[9] Development work has also been divided into independent modules out of which hacker groups create competing versions. A group consisting of Torvalds and a few other principal developers then decides which of these versions will be incorporated in the improved version of Linux (and, of course, the modular structure also develops gradually). Torvalds's group does not, however, hold any permanent position of authority. The group retains its authority only for as long as its choices correspond with the considered choices of the hacker community. Should the group's choice prove less than enlightened, the hacker community proceeds to develop the project in its own direction, bypassing the former leaders of the pack.

In order to control the continuous development of Linux, publications have been divided into two series. In the stable versions, safe for use by average users, the *y* in the release number *x.y.z* is even (e.g., version 1.0.0), whereas in the developmental versions, aimed at programmers, the *y* is the stable version's $y + 1$ (e.g., the stable version 1.0.0's improved but still not finally tested developmental version is 1.1.0). *X* grows only when a truly fundamental change is made (at the time of writing, the latest available version is 2.4.0). This simple model has worked surprisingly well in the management of Linux development.

In the well-known essay "The Cathedral and the Bazaar," published originally on the Net, Raymond has defined the difference between Linux's open model and the closed model preferred by most companies by comparing them to the bazaar and the cathedral. Although a technologist himself, Raymond emphasizes that Linux's real innovation was not technical but social: it was the new, completely open social manner in which it was developed. In his vocabulary, it was the shift from the cathedral to the bazaar.[10]

Raymond defines the cathedral as a model in which one

person or a very small group of people plans everything in advance and then realizes the plan under its own power. Development occurs behind closed doors, and everybody else will see only the "finished" results. In the bazaar model, on the other hand, ideation is open to everyone, and ideas are handed out to be tested by others from the very beginning. The multiplicity of viewpoints is important: when ideas are disseminated widely in an early stage, they can still benefit from external additions and criticisms by others, whereas when a cathedral is presented in its finished form, its foundations can no longer be changed. In the bazaar, people try out different approaches, and, when someone has a brilliant idea, the others adopt it and build upon it.

Generally speaking, this open-source model can be described as follows: it all begins with a problem or goal someone finds personally significant. That person may release just the problem or goal itself, but usually he or she will also provide a Solution—version 0.1.1, to use the Linux numbering system. In the open model, a recipient has the right to freely use, test, and develop this Solution. This is possible only if the information that has led to the Solution (the source) has been passed on with it. In the open-source model, the release of these rights entails two obligations: these same rights have to be passed on when the original Solution or its refined version (0.1.2) is shared, and the contributors must always be credited whenever either version is shared. All this is a shared process, in which the participants move gradually—or sometimes even by leaps and bounds (say, a shift from version 0.*y.z* to version 1.*y.z*)—to better versions. In practice, of course, projects follow this idealized model to a greater or lesser extent.

The Academy and the Monastery

Another possible allegory for the open-source model is again the academy, which it resembles even more directly than the cathedral. Scientists, too, release their work openly to others for their use, testing, and further development. Their research is based on the idea of an open and self-correcting process. The latter idea of self-correction was emphasized by Robert Merton as an equally important cornerstone of scientific ethic as openness. He called it *organized skepticism*[11]—historically, it is a continuation of the *synusia* of Plato's Academy, which also included the idea of approaching the truth through critical dialogue.[12] The scientific ethic entails a model in which theories are developed collectively and their flaws are perceived and gradually removed by means of criticism provided by the entire scientific community.[13]

Of course, scientists, too, have chosen this model not only for ethical reasons but also because it has proved to be the most successful way of creating scientific knowledge. All of our understanding of nature is based on this academic or scientific model. The reason why the original hackers' open-source model works so effectively seems to be—in addition to the facts that they are realizing their passions and are motivated by peer recognition, as scientists are also—that to a great degree it conforms to the ideal open academic model, which is historically the best adapted for information creation.

Broadly speaking, one can say that in the academic model the point of departure also tends to be a problem or goal researchers find personally interesting; they then provide their own Solution (even though in many instances the mere statement of the problem or proclamation of a program is interesting in itself). The academic ethic demands that anyone may use, criticize, and develop this Solution. More important than any final result is the underlying information or chain of arguments that has produced the Solution. (It is not enough to merely publish "$E = mc^2$"—theoretical and empirical justifications are also required.) Nevertheless, the scientific ethic does not involve only rights; it also has the same two fundamental obligations: the sources must always be mentioned (plagiarism is ethically abhorrent), and the new Solution must not be kept secret but must be published again for the benefit of the scientific community. The fulfillment of these two obligations is not required by law but by the scientific community's internal, powerful moral sanctions.

Following this model, normal physics research, for example, continuously provides new additions ("developmental versions") to what has already been achieved, and after testing these refinements the scientific community accepts them as part of its body of knowledge ("stable versions"). Much more rarely, there is an entire "paradigm shift," to use the expression that philosopher of science Thomas Kuhn introduced in his book *The Structure of Scientific Revolutions*.[14] In the broadest sense, there have been only three long-lived research paradigms in physics: the Aristotelian-Ptolemaic physics, the "classic" Newtonian physics, and the Einsteinian-Heisenbergian physics based on the theory of relativity and quantum mechanics. Seen this way, present theories are versions 3.*y*.*z*. (Many physicists already call the version 4, which they believe is imminent, "The Theory of Everything." Computer hackers would not anticipate the arrival of version 4.0.0 quite so eagerly.)

The opposite of this hacker and academic open model can be called the closed model, which does not just close off information but is also authoritarian. In a business enterprise built on the monastery model, authority sets the goal and chooses a closed group of people to implement it. After the group has completed its own testing, others have to accept the result as it is. Other uses of it are "unauthorized uses." We can again use our allegory of the monastery as an apt metaphor for this style, which is well summed up by Saint Basil the Great's monastic rule from the fourth century: "No one is to concern himself with the superior's method of administration."[15] The closed model does not allow for initiative or criticism that would enable an activity to become more creative and self-corrective.

We have mentioned that hackers oppose hierarchical operation for such ethical reasons as that it easily leads to a culture in which people are humiliated, but they also think that the nonhierarchical manner is the most effective one. From the point of view of a traditionally structured business, this may initially seem quite senseless. How could it ever work? Should not someone draw an organization chart for the Net and Linux developers? It is interesting to note that similar things might be said of science. How could Einstein ever arrive at his $E = mc^2$ in the chaos of self-organized groups of researchers? Should science not operate with a clear-cut hierarchy, headed up by a CEO of Science, with a division chief for every discipline?

Both scientists and hackers have learned from experience that the lack of strong structures is one of the reasons why this model is so powerful. Hackers and scientists can just start to realize their passions and then network with other individuals who share them. This spirit clearly differs from that found not only in business but also in government. In governmental agencies, the idea of authority permeates an action even more strongly than it does in companies. For the hackers, the typical governmental way of having endless meetings, forming countless committees, drafting tedious strategy papers, and so on before anything happens is at least as great a pain as doing market research to justify an idea before you can start to create. (It also irritates scientists and hackers no end when the university is turned into a governmental bureaucracy or monastery.)

But the relative lack of structures does not mean that

there are no structures. Despite its apparent tumult, hackerism does not exist in a state of anarchy any more than science does. Hacker and scientific projects have their relative guiding figures, such as Torvalds, whose task it is to help in determining direction and supporting the creativity of others. In addition, both the academic and hacker models have a special publication structure. Research is open to anyone, but in practice contributions included in reputable scientific publications are selected by a smaller group of referees. Still, this model is designed so as to guarantee that, in the long run, it is the truth that determines the referee group rather than the other way around. Like the academic referee group, the hacker network's referee group retains its position only as long as its choices correspond to the considered choices of the entire peer community. If the referee group is unable to do this, the community bypasses it and creates new channels. This means that at the bottom the authority status is open to anyone and is based only on achievement—no one can achieve permanent tenure. No one can assume a position in which his or her work could not be reviewed by peers, just as anyone else's creations can be.

The Hacker Learning Model

It goes without saying that the academy was very influential long before there were computer hackers. For example, from the nineteenth century onward, every industrial technology (electricity, telephone, television, etc.) would have been unthinkable without its underpinning of scientific theory. The late industrial revolution already marked a transition to a society that relied upon scientific results; the hackers bring about a reminder that, in the information age, even more important than discrete scientific results is the *open academic model* that enables the creation of these results.

This is a central insight. In fact, it is so important that the second big reason for the pragmatic success of the hacker model seems to be the fact that hackers' learning is modeled the same way as their development of new software (which can actually be seen as the frontier of their collective learning). Thus, their learning model has the same strengths as the development model.

A typical hacker's learning process starts out with setting up an interesting problem, working toward a solution by using various sources, then submitting the solution to extensive testing. Learning more about a subject becomes the hacker's passion. Linus Torvalds initially taught himself programming on a computer he inherited from his grandfather. He set up problems for himself and found out what he needed to know to solve them. Many hackers have learned programming in a similarly informal way, following their passions. The example of the ability of ten-year-olds to learn very complicated programming issues tells us much about the importance of passion in the learning process, as opposed to the slow going their contemporaries often find their education in traditional schools to be.[16]

Later on, the beginnings of Torvalds's operating system arose out of his explorations into the processor of the PC he purchased in 1991. In typical hacker fashion, simple experiments with a program that just tested the features of the processor by writing out either *A*s or *B*s gradually expanded into a plan for a Net newsgroup-reading program and then on to the ambitious idea of an entire operating system.[17] But even though Torvalds is a self-taught programmer in the sense that he acquired his basic knowledge without taking a class, he did not learn everything all by himself. For example, in order to familiarize himself with operating systems, he studied the source codes of Tanenbaum's Minix as well as various other information sources provided by the hacker community. From the very beginning, in true hacker fashion, he has never hesitated to ask for help with questions in areas in which he has not yet acquired expertise.

A prime strength of the hacker learning model lies in the fact that a hacker's learning teaches others. When a hacker studies the source code of a program, he often develops it further, and others can learn from this work. When a hacker checks out information sources maintained on the Net, he often adds helpful information from his own experience. An ongoing, critical, evolutionary discussion forms around various problems. The reward for participating in this discussion is peer recognition.

The hackers' open learning model can be called their "Net Academy." It is a continuously evolving learning environment created by the learners themselves. The learning model adopted by hackers has many advantages. In the hacker world, the teachers or assemblers of information sources are often those who have just learned something. This is beneficial because often someone just engaged in the study of a subject is better able to teach it to others than the expert who no longer comes to it fresh and has, in a way, already lost his grasp of how novices think. For an expert, empathizing with someone who is just learning something involves levels of simplification that he or she often resists for intellectual reasons. Nor does the expert necessarily find the teaching of basics very satisfying, while a student may find doing such teaching tremendously rewarding, since he or she does not as a rule get to enjoy the position of instructor and is generally not given sufficient opportunity to use his or her talents. The process of teaching also involves by its very nature the comprehensive analysis of subject matter. If one is really able to teach something to others, one must have already made the material very clear to oneself. While preparing the material, one has to consider it carefully from the point of view of possible further questions and counterarguments.

Once again, this hacker model resembles Plato's Academy, where students were not regarded as targets for knowledge transmission but were referred to as companions in learning (*synetheis*).[18] In the Academy's view, the central task of teaching was to strengthen the learners' ability to pose problems, develop lines of thought, and present criticism. As a result, the teacher was metaphorically referred to as a midwife,[19] a matchmaker,[20] and a master of ceremonies at banquets.[21] It was not the teacher's task to inculcate the students with preestablished knowledge but to help them give birth to things from their own starting points.

In the hacker community, too, the experts understand themselves as learners who can just act as gadflies, midwives, and symposiarchs to others, thanks to their deeper knowledge.

The Net Academy

The ethos of the original academic and the hacker model—well summed up by Plato's idea that "no free person should learn anything like a slave"[22]—is totally different from that of the monastery (school), the spirit of which was summed up by Benedict's monastic rule: "It belongeth to the master to speak and to teach; it becometh the disciple to be silent and to listen."[23] The irony is that currently the academy tends to model its learning structure on the monastic sender-receiver model. The irony is usually only amplified when the academy starts to build a "virtual university": the result is a computerized monastery school.

The scientific revolution in the seventeenth century was supposed to mean the abandonment of scholasticism and its replacement with a science continually striving for new knowledge. Nevertheless, the university has preserved the scholastic teaching model and hierarchy, down to its vocabulary (e.g., a "dean" was originally an office-holder of a monastery). The scientific revolution took place four hundred years ago, but it is not very well reflected in our universities as a basis for research-based learning. It seems quite strange that we expect scholastic teaching methods to be able to produce modern individuals capable of independent thought and the creation of new knowledge.

The wider significance of the hacker learning model is its healthy reminder to us of the potential in the original idea of seeing the academic development and learning models as identical. We could also use this idea to create a generalized Net Academy, in which all study materials would be free for use, critique, and development by everyone. By improving existing material in new directions, the network would continuously produce better resources for the study of the subjects at hand. Members of the network would be driven by their passions for various subjects and by the peer recognition for their contributions.

Logically, the continued expansion and development of this material, as well as the discussion and examination of it, would also have to be the Net Academy's only way to grant study credits; and, true to the spirit, the highest credits should be given for those accomplishments that prove the most valuable to the entire learning community. A hacker-style reading of the material with a view toward criticizing and improving it—that is, toward doing something, motivating oneself, with it—would also be much more conducive to learning than the current tendency to just read material.

The Net Academy would follow the hacker model in creating an important continuum from the beginning student all the way to the foremost researcher in the field. Students would learn by becoming researching learners from the very beginning, by discussing matters with researchers, and later on by studying the research publications of their field directly.

In the Net Academy, every learning event would permanently enrich all other learners. Alone or in the company of others, the learner would add something to the shared material. This differs from our present mode of disposable learning, in which every student starts from the beginning, passes the same exams isolated from everyone else, and never gets to benefit from the insights of others. Worse, after the exam the examiner basically tosses all those individual insights into the wastebasket. This is as absurd a procedure as would be the decision of each generation of researchers to finally toss all their results away ("I see, $E = mc^2$; so what—toss!") and let the next generation start over.[24]

It goes without saying that the practical realization of

the general Net Academy presents a great challenge. For example, as in the world of hackers and researchers, a guiding structure for the collective creation of learning materials is needed. When material is constantly adapted and expanded in new directions, competing versions are born. This is always the case in the hacker and research fields. Hackers have solved practical problems arising from this by developing so-called concurrent-versioning systems: these enable one to see how competing versions differ from the existing version and from each other. On a more theoretical level, the problem can be solved by the practice of referees. With the help of a concurrent-versioning system, a self-organized group of referees can make decisions between competing versions and combine their ideas if need be.

After the hackers' reminder of the full significance of the academic model, it would be odd to continue our current practice of providing learners mainly with results, without making them learn much more deeply the academic model itself, which is based on a collective process of posing of problems, the questioning of them, and the development of solutions—a process driven by passion and recognition for socially valuable contributions. The core of the academy does not consist of its individual achievements but of the academic model itself.

The Social Model

Expressing this one possible wider application inherent in the hacker model must not, of course, be understood to say that we should just wait for governments or corporations to execute it. A central point of hackerism is to remind us that through the open model great things can be accomplished by individuals' direct cooperation. The only limit is our imagination. For example, the hacker open model could be transformed into a social model—call it the open-resource model—in which someone announces: I have an idea, I can contribute this much to it, please join me! Although this version of the open model would also involve local physical action, the Net would be used as an effective means for joining forces and later disseminating and developing the idea further.

For example, I could announce on the Net that I would be willing, once in a while, to help some elderly person take care of things. I can announce that kids can come and play at our house after school. I can say that I would be glad to walk one of the neighborhood dogs on weekdays. Perhaps the effectiveness of this model could be strengthened by adding a condition that the helped person commit to helping someone else equally. The Net can be used as a means to organize local resources. Gradually, others will join the realization of great social ideas, and this will generate even greater ideas. There would be a self-feeding effect, as in the computer hacker model.

We have seen that the hacker model can bring about great things in cyberspace without governments and corporations as mediators. It remains to be seen what great things individuals' direct cooperation will accomplish in our "flesh reality."

PART THREE

The NETHIC

CHAPTER 5

From Netiquette to a Nethic

Netiquette and Nethic

Beyond the hacker work and money ethic is the significant third level of the hacker ethic that can be called the *nethic* or network ethic. This expression refers to the hackers' relationship to our network society's networks in a wider sense than the more familiar term *netiquette* (which concerns behavioral principles for communication on the Net—e.g., "avoid flaming," "read the file of frequently asked questions before posting your message," etc.).[1] Again, not all hackers share all the elements of the nethic, but still these elements are linked together in their social meaning and relation to the hacker ethic.

The first part of the hacker nethic consists of the hackers' relation to media networks such as the Net. Although we can say that hackers' characteristic relation to them dates back to the origin of the hacker ethic in the sixties, this nethic has received a more conscious formulation in recent years. One key moment came in 1990 when hackers Mitch Kapor and John Perry Barlow started the Electronic Frontier Foundation in San Francisco to promote the fundamental rights of cyberspace.[2] Barlow, a child of the sixties counterculture, used to write songs for the Grateful Dead and became a pioneer in the cyber-rights movement. He was the first to apply William Gibson's term *cyberspace* (from his novel *Neuromancer*) to all electronic networks.[3] Kapor was an important player in the development of personal computers, creating, in 1982, the spreadsheet program Lotus. It was the first PC application that made a widespread function significantly easier than it had been before, and this made it an important factor in the breakthrough of the personal computer.[4] The name *Lotus* reflected Kapor's background: as a former mental-health counselor with a psychology degree, and later a transcendental-meditation instructor, he was interested in Eastern thought systems.

The enterprise, also called Lotus, that Kapor had built around his program quickly evolved into the largest software company of its time. But as his original hackerism became more and more entrepreneurial, Kapor began to feel alienated, and he left the business after four years. In his own words: "It felt awful to me, personally. So I left. I just walked away one day. . . . The things that were important to the business as an organism were things that I could demonstrate less and less enthusiasm for."[5]

Both Barlow and Kapor considered the fundamental rights of cyberspace, such as freedom of speech and privacy, critical issues. The immediate impetus for the Electronic Frontier Foundation was the FBI's suspicion that both Barlow and Kapor were in the possession of stolen source code. So, in popular usage, they were suspected of being "hackers" (that is, crackers), and FBI agents paid visits to both of them. The suspicion was unfounded, but Barlow and Kapor felt that lawmakers and law enforcers did not really understand what genuine hackerism and cyberspace were all about. For example, the agent visiting Barlow hardly knew anything about computers and referred to Nu Prometheus, the cracker group that had stolen the code, as New Prosthesis.

Barlow and Kapor could have shrugged off those visits, but they became concerned that a lack of understanding might ultimately lead to a totalitarian-style regulation of electronic space that could seriously weaken the freedom of speech and privacy dear to hackers. Ironically, Barlow's FBI visitor—a defender of capitalist law and order—happened to be a namesake of the Protestant preacher Richard Baxter that Weber saw as the purest representative of the Protestant ethic, almost as if the meeting had been scripted as an allegorical confrontation between the Protestant ethic and the hacker ethic.

The EFF's cofounders included Wozniak, John Gilmore, and Stewart Brand. Gilmore is known for supporting the use of strong-encryption technologies to protect privacy, and for his slogan "the Net treats censorship as damage and routes around it," in accordance with which he cofounded the Net's totally uncontrolled alt- newsgroups. Brand was the creator of *The Whole Earth Catalog* and played a significant role in the history of hackerism by writing the first article about it (in *Rolling Stone* in 1972) and by organizing the first Hacker Conference (in San Francisco in 1984).

The EFF defines itself as "a non-profit, non-partisan organization working in the public interest to protect fundamental civil liberties, including privacy and freedom of expression, in the arena of computers and the Internet."[6] In practice, the EFF has contributed to the overturning of, among other measures, the Communication Decency Act passed by the United States Congress in 1997, which tried to create a kind of censorship authority for the Internet. The EFF has also played an important part in defending the use of strong-encryption technologies previously declared illegal in the United States. Before this law was changed, the EFF, through Gilmore, constructed the DES Cracker, which was capable of breaking through the so-called DES protection used in the encryption of some bank transactions and e-mail delivery on the Net; the point was to demonstrate that the encryption methods permitted by the United States are not able to protect privacy.[7] Socially conscious hackers emphasize that encryption technology must not only meet the encryption needs of governments and businesses but also protect the individual from governments and businesses.

Freedom of expression and privacy have been important hacker ideals, and the Net developed in accordance with them. The need for hacker organizations such as the EFF arose when, in the nineties, governments and enterprises became interested in the Net on a large scale and have since often tried to develop it in a direction opposed to hacker ideals.

In its defense of freedom of expression and privacy, the hacker world is typically decentralized. In addition to the EFF, there is a great number of other hacker groups engaged in similar activity. Two examples of these are Holland's ethically engaged XS4ALL Internet service and Witness, which reports on crimes against humanity using the tools of cyberspace. These hacker groups join forces in thematic clusters such as the Global Internet Liberty Campaign.[8]

Freedom of Speech: The Case of Kosovo

There is more than enough for all of them to do. In the so-called developed countries, where freedom of expression and privacy are considered fundamental rights, there are nevertheless continued attempts to curtail these rights in cyberspace.[9] In the rest of the world, however, these rights are not even recognized in any clear fashion. According to *Censor Dot Gov: The Internet and Press Freedom 2000,* a study published by the research center Freedom House, about two thirds of the world's countries and four fifths of the world's population do not have complete freedom of speech as of the beginning of 2000.[10]

When so minded, the powers that be are able to control the media, especially the traditional, centralized ones such as the press, radio, and television. They do, of course, also try to gain control over Net content, but in practice this is remarkably difficult, due to the Net's decentralized structure. For this reason, the Net has become an important medium for free individual expression in totalitarian societies. And hackers, who created this medium, from e-mail and newsgroups to chat and the Web, have helped dissidents in various parts of the world in its use.

The Kosovo crisis of 1999 is an excellent example of these attempts, seen in many other countries too.[11] Censorship is often an early-warning symptom of other forthcoming violations of human rights, and once these violations are being carried out, censorship allows for only a sanitized official version of events and prevents the dissemination of any criticism. This was the case in Yugoslavia, where President Slobodan Milosevic gradually tightened his stranglehold on the media while the country's Serb majority speeded up "ethnic cleansing" in the province of Kosovo, the Albanian majority of which wanted self-government.

Things get ugly when freedom of speech is curtailed.

While Serb forces in Kosovo executed men, raped women, and drove entire villages—from newborn children to the aged—into exile, the official media of Yugoslavia proclaimed that everything was just fine. (This tradition was continued until the last moments of Milosevic's power: after he had tampered with election results and while hundreds of thousands of people were protesting in the center of Belgrade, Serbian TV showed Olympics replays and classical music.) The media could not report on atrocities, and opponents' voices were silenced. During the NATO air strikes aimed at putting an end to the massacres, the traditional Yugoslavian media were practically taken over by the government. The academy was also silenced because it is the traditional proponent of free speech.[12] The words of Basil's rule, "No one is to . . . make curious inquiries about what is being done," described the government's policy.

The Net, however, was able to spread the news. On the initiative of the EFF, a network server named anonymizer.com provided Kosovars with the opportunity to send out messages in a way that prevented authorities from tracking them down.[13] However, the best-known messages from the war were transmitted as straightforward e-mail. A famous instance was the e-mail correspondence between "Adona," a sixteen-year-old ethnic Albanian girl, and Finnegan Hamill, a junior at Berkeley High School in California. (Adona's real identity was not revealed for security reasons.) Adona wrote:

> Hello Finnegan. . . . One night, last week I think, we were all surrounded by police and armed forces, and if it wasn't for the OSCE observers, God knows how many victims there would be. And my flat was surrounded too. I cannot describe you the fear. . . . The next day, a few meters from my flat, they killed this Albanian journalist, Enver Maloku. Someday before there was a bomb explosion in the center of town where young people usually go out.[14]

On another day she wrote:

> I don't even know how many people get killed anymore. You just see them in the memoriam pages of newspapers. I really don't want to end up raped, with no parts of the body like the massacred ones. I wish nobody in the world, in the whole universe would have to go through what we are. You don't know how lucky you are to have a normal life. We all want to be free and living like you do, having our rights and not be pushed and pushed. Finnegan, I'm telling you how I feel about this war and my friends feel the same.

Just before the beginning of the NATO air strikes, Adona sent this message:

> Dear Finnie,
> At the moment I am writing to you, just from my balcony. I can see people running with suitcases and I can hear some gunshots. A village just a few meters from my home is all surrounded. I have prepared my

> bag with necessary things: clothes, documents, and money . . . in case of emergency. Only the past few days, there have been so many new forces, tanks, and soldiers coming inside Kosovo. Yesterday, a part of my town was surrounded and there were shootings happening. . . . I am waiting with no patience for the news.

The control exercised by the Milosevic government was based both on the strict "public information law" of 1998, which permitted the closing down of the media on the whim of the authorities, and on sheer brute force. For example, in March 1999, Serb police shot and killed Bajram Kelmendi, a human-rights advocate, and his two sons. Kelmendi had been defending the Albanian-language newspaper the police had closed down. Slavko Curuvija, publisher of two independent newspapers and a man who according to government television supported the NATO air strikes, was shot down in front of his home on April 11, 1999. Dozens of other journalists were arrested, brutalized, or sent into exile.[15]

Yugoslavia's most influential oppositional medium, the radio station B92, has had continuous and varied problems with the authorities. On November 27, 1996, during antigovernment demonstrations, its transmission signal was jammed, and on December 3 it was shut down completely. At this point, XS4ALL offered to help B92 by relaying its transmissions over the Net (the sound-transmission technology was provided by RealAudio from RealNetworks, which is financed by Kapor). The Voice of

America, among others, then transmitted the signal received through the Net back to Yugoslavia. Its censorship having proved ineffective, the government soon allowed B92 to resume normal radio transmissions.[16]

XS4ALL's ideology is expressed in its name: Net access should be available to all, as the Net is a medium for freedom of expression. XS4ALL says it is ready to be "active in politics and is not frightened of lawsuits."[17] Cooperation between XS4ALL and B92 resumed at the beginning of the Kosovo war on March 24, 1999, when Yugoslavia's telecommunications ministry once again closed down the station and confiscated its transmitters. The station's editor in chief, Veran Matic, was arrested, though released the same day, with no explanation given. On April 2, the station's director, Sasa Mirkovic, was fired, and the authorities appointed a new director and prescribed new guidelines. With the help of XS4ALL, the original editors of B92 managed to continue their transmissions, again via the Net, and radio stations abroad once more transmitted the signal back to Yugoslavia.[18]

B92's victory over government control was particularly important in that the station became a symbol for independent critical media in Yugoslavia. The defense of free media written by Matic at the beginning of the war expresses well what was at stake: "As a representative of the free media, I am only too aware of the need for information, whatever side you are on in the conflict. People inside the country should be kept up to date with international debate as well as with what is happening at home. Those abroad ought to be told the truth about what is going on here. But instead of detailed, uncensored facts, all we hear is war propaganda, including western rhetoric."

Toward the end of the war, the Witness organization trained four Kosovars to document human-rights abuses on digital video. The visual material was then transmitted out of the country by means of a laptop computer and satellite phone via the Net. This material has been made available to the International War Crimes Tribunal.[19]

Witness, founded in 1992, believes in the power of images in the reporting of human-rights violations and defines its task as the development of video technology and training in its use for this purpose: "Our goal is to equip human rights defenders with the tools they need to record, transmit, and publicize human rights abuses that might otherwise go unnoticed and unpunished." Its founder, musician and cyberart pioneer Peter Gabriel puts it: "Truth knows no borders. Information wants to be free. Technology is the key."[20]

In addition to these hacker groups, even the more traditional action groups moved into "Net time" during the Kosovo conflict. OneWorld, which coordinates civilian organizations, and its partner Out There News created a Net database of refugees, to help people find their relatives and friends.[21] Even in the peace-treaty negotiations, which were naturally determined above all by human and not technological factors, the new technology played a symbolic part. In the negotiations chaired by Finland's president Martti Ahtisaari and Russia's former prime minister Viktor Chernomyrdin, the first draft treaty was written on a Net mobile phone, and the first preliminary reports on the negotiations were sent to representatives of different countries as text messages.[22] Thus, it may be justified to call the Kosovo war the first Net war, the way the Vietnam War has been labeled the first television war.

A small part of the war was even conducted over the Net. Crackers supporting different sides launched their own attacks, as described in Dorothy E. Denning's study *Activism, Hacktivism, and Cyberterrorism* (2000). Serbian crackers disrupted the NATO server only a couple of days after the war began. A Californian cracker countered with an attack on the Yugoslav government's webpages. Crackers took sides according to their views of the conflict: the Russians and the Chinese attacked the United States, and Americans, Albanians, and western Europeans attacked Serb pages. Some eastern European crackers also created viruses with anti-NATO messages. After the end of the war, some media even spread the (false) rumor that President Clinton had approved of a plan to use crackers for operations such as the raiding of Milosevic's bank accounts.[23]

It must be admitted that the Net had only minor influence on general views of the war, and even less on its conduct. Nevertheless, there is no reason to consider it, as a medium for free speech, as something separate from other media, since all media are interconnected in their spheres of influence. As a reception channel, the Net still is not a mass medium, but that statement demands two important qualifications. First, in some circumstances the Net may be irreplaceable as a reception channel. Via it, messages from the traditional media may reach audiences that have been prevented from receiving them by censorship exercised by their own governments. This is how many people in totalitarian countries receive information and views that are not allowed by their governments.

Second, the Net does not necessarily have to be a mass medium reception channel to have an influence on a wide public. It can be an effective production tool in the creation of reports that can then be disseminated through the traditional media. We must remember that the Net provides everyone with the tools of a journalist. Even the reporters and editors employed by the traditional media increasingly write, record video for, and transmit their stories using these tools. When computers, telecommunications, and the traditional media combine forces in one fast multimedia Net, and when the computer, the telephone, and the camera are fused into a small multimedia gadget, individuals become able to transmit reports designed for the great media machinery. Such a user of future Net appliances may not operate on a technical and journalistic level comparable to that of professionals, but these shortcomings are outweighed by his or her being on the spot and experiencing events firsthand. In Kosovo, we have seen only the beginning of what media hackerism may achieve.

Privacy or Electronic Omniscience

The Net may be a medium of free speech, but it can also be turned into a medium of surveillance. Traditionally, many hackers have worked to prevent this by also defending privacy in cyberspace. Recently, governments and business have tried to make inroads on this privacy in many ways.[24]

In a number of countries, there have been discussions about a so-called back door to the Net, which governments could use for surveillance when they consider it necessary, or even automatically to keep a permanent eye on people's e-mail and Web-browsing patterns. (Automated surveillance is based on programs that analyze the contents of messages and Web visits and report "dubious" cases to a human surveillance agent.) In this respect, the difference between so-called developed and developing countries seems to be that there still is a debate about these tactics in the developed countries, while in the developing countries governments already use such devices without any preliminary discussions. Thus, in Saudi Arabia Internet-service providers are obliged to keep a log of users' activity on the Web and to send an automated warning to users as soon as they try to access banned sites or pages, to remind them that they are, indeed, being watched.[25]

In developed countries, at least in peacetime, business is a greater menace to privacy than the government tends to be. Although business enterprises cannot access Internet-service providers' databases in the manner that governments can, they are able to deduce similar information in other ways. While moving around the Web, a user's browser program and webpage servers exchange information that identifies the user (so-called cookies). By itself, this does not enable anyone to know the user's personal data, but it does allow them to note each time a user *x* visits a particular webpage. After this, the identification of *x* can be made, at least in principle, as soon as the person gives out personal information to any website that collects such information and sells it to others. After that, *x* has a name, a sex, an age, an address, an e-mail address, and so on. And after that, it can be known who visits dog pages, pages concerning a certain pop artist, pornographic pages, et cetera, on the basis of which a person's interests can be analyzed.

Some enterprises have specialized in the collection of such information by placing advertisements on a huge number of webpages. Since these advertisements are not really part of the page but are provided by the advertiser's Web server, the advertiser is also able to exchange identifying information with the user's browser. The main purpose of these advertisements—or, more accurately, "spy links"—is to gather information on the browsing patterns of individuals. The lifestyles of individuals are the stock in trade of these enterprises. The comprehensiveness of the lifestyle maps they generate from this information depends on how many spy pages the enterprise is able to maintain and how much information on their visitors or clients businesses outside of its spy ring are willing to sell to them.

Messages sent to newsgroups are another essential source of lifestyle information. They are easier to analyze, since basically all messages sent to newsgroups are saved permanently somewhere, open for anyone to read. Surprising amounts of information can be gathered by simply observing what groups individuals have participated in and by analyzing the language of their messages.

In the electronic era, users constantly leave electronic traces in various databases. The more electronic our era becomes, the more traces there are to be found. Thus, as computers, telephones, and the media converge, even the television programs people watch, the radio stations they tune in to in their cars, and the articles they read in the online newspapers can be recorded in electronic databases. Through the base stations mobile-phone owners use, even their locations can be determined with extreme accuracy. With this kind of information, a very intimate image of an individual can be created.

As the number of electronic traces increases, the image of the individual becomes more and more accurate. Even today, every bank and credit-card transaction is recorded in the card company's database; if a person uses a courtesy card, transactions made with it are also recorded in that company's database. The electronic currency of the future (whether it will be used via the computer, the mobile phone, the television set, or some other appliance) will preserve this information even more comprehensively. In the most detailed case, some database might list each and every product bought by individuals during their lifetimes. It is easy to see how a detailed profile of a person could thus be created.

Specific knowledge of an individual's lifestyle interests business enterprises for two main reasons. First, such knowledge facilitates very precisely aimed marketing: for example, when a person is known to have a dog, she or he will receive commercials for dog products on her or his digital TV during commercial breaks. (Had this person also once sent an e-mail message with the title "Cats suck," he or she would also not receive commercials for cat products.) Or if it is known that a person has a sweet tooth, she or he could receive, at appropriate moments during the day, mobile-phone messages about candy offerings at a nearby store.

Second, such detailed profiling makes it possible to scrutinize the lifestyles of workers and job applicants. The storage of people's doings in electronic memory means, ultimately, that no act remains unknown. In the electronic age, the corporate monastery's gate is guarded by a computerized Saint Peter, who differs from omniscient God only in that he is not forgiving. During the job interview, the applicant's entire life up to that moment flashes by, and he or she has to account for all sins: at age six, you flamed your buddy on the Net in a politically incorrect manner; at fourteen, you visited pornographic websites; at eighteen, you confessed to a chat room that you had experimented with drugs. . . .

An increasing number of businesses also exercise (sometimes unannounced) surveillance over their employees' electronic behavior. Many enterprises have installed computer programs that observe employees' use of e-mail and the Web: does the employee use inappropriate language (e.g., expressions of anger); with whom is he or she in touch (not the competitor, we hope); does he or she visit sites of ill repute (those pornography pages)? Even the content of telephone conversations can be controlled in a similar manner using speech-to-text technology.[26]

Hackers have long emphasized that the maintenance of privacy is by no means a given in the electronic age but requires more conscious protection than ever before. They have spent much time discussing the pressures on privacy now exercised by businesses and governments. For the sake of privacy, some hackers have even resorted, symbolically, to preelectronic solutions in some particularly intrusive circumstances. Eric Raymond, for instance, does not use a bank card, because he is opposed to the way in which the technical operation of that system records every monetary transaction. Technically, it would have been possible to create a model in which the individual's transactions would not transmit any personal information and the business would still be able to charge the correct card. This is a matter of choice.

Many hackers abhor any violation of the individual's personal boundaries, no matter whether this takes place during working hours or outside of them. An employment relationship gives no one the right to intrude into personal territory. Danny Hillis's Zen-like anecdote about personality testing exemplifies the way hackers feel about employers' eagerness to analyze the individual ever more precisely, using all kinds of techniques: "A disciple of another sect once came to Drescher [a researcher at Minsky's AI Lab] as he was eating his morning meal. 'I would like to give you this personality test,' said the outsider, 'because I want you to be happy.' Drescher took the paper that was offered him and put it into the toaster, saying: 'I wish the toaster to be happy, too.' "[27]

In order to protect electronic privacy, many hackers have defended the use of the kinds of strong-encryption technology of which governments disapprove, since strong encryption is needed to guarantee genuine privacy. The U.S. law on arms exports formerly classified these technologies (which use a key larger than 64 bits) as munitions, and thus their sales came under strict regulation. To parody this law, one hacker tattooed on his left arm the so-called RSA encryption method, classified as strong encryption, in just three short lines of code, which he accompanied, in compliance with U.S. law, with this statement: WARNING: THIS MAN IS CLASSIFIED AS A MUNITION.

FEDERAL LAW PROHIBITS TRANSFER OF THIS MAN TO FOREIGNERS.[28]

Hacker groups played an important role in bringing about some loosening of these legal restrictions in early 2000.[29] One of the most important groups developing strong-encryption methods is Cypherpunks, founded by John Gilmore, Tim May, and Eric Hughes. Its goals are summed up in Hughes's "A Cypherpunk's Manifesto" of 1993:

> We must defend our own privacy if we expect to have any. We must come together and create systems which allow anonymous transactions to take place. People have been defending their own privacy for centuries with whispers, darkness, envelopes, closed doors, secret handshakes, and couriers. The technologies of the past did not allow for strong privacy, but electronic technologies do.
>
> We the Cypherpunks are dedicated to building anonymous systems. We are defending our privacy with cryptography, with anonymous mail forwarding systems, with digital signatures, and with electronic money.[30]

In his manifesto "Privacy, Technology, and the Open Society" (1991), John Gilmore fantasizes further about what a society constructed on hacker principles might be like:

> What if we could build a society where the information was never collected? Where you could pay to rent a video without leaving a credit card number or a bank number? Where you could prove you're certified to drive without ever giving your name? Where you could send and receive messages without revealing your physical location, like an electronic post office box?
>
> That's the kind of society I want to build.[31]

Hackers work to find technical solutions that will enable the electronic age to respect privacy. The Cypherpunks are by no means alone in realizing this ambition. The first functional anonymous server that enabled people to send e-mail or messages to newsgroups without revealing their identities (known as a remailer) was created by a Finnish hacker, Johan Helsingius. A member of Finland's Swedish-speaking minority himself, he describes the need for such a server: "Where you're dealing with minorities—racial, political, sexual, whatever—you always find cases in which people belonging to a minority would like to discuss things that are important to them without having to identify who they are." In another context, he adds: "These remailers have made it possible for people to discuss very sensitive matters, such as domestic violence, school bullying or human rights issues anonymously and confidentially on the Internet."[32]

In the future, privacy won't be merely an ethical question but a technological one as well. The technical realization of electronic networks has a great impact on the individual's right to privacy. The hacker nethic's defense of privacy becomes a hard cooperative effort: besides securing the Net, influence has to be exerted on a great number of other networks that store details of individuals' lives.

Virtual Reality

Historically, the Net as a hacker medium has an important third dimension that is not often linked to the idea of the hacker ethic, although it is clearly related to the above two attitudes toward the media: in addition to the ideas of freedom of expression and privacy, hackers value the individual's own activity. In fact, *activity* is a word that sums up well the linking idea behind all three elements of the hackers' nethic. Freedom of expression is a means toward being a publicly active member of society, receiving and articulating various views. Privacy secures one's activity in creating a personal lifestyle, because surveillance is used in order to persuade people to live in certain ways or to deny legitimacy to lifestyles that deviate from the ruling norms. Self-activity emphasizes the realization of a person's passion instead of encouraging a person to be just a passive receiver in life.

In this last respect, the nature of traditional media (especially television), which makes the user merely a receiver, is very different. It takes the monastic idea of a one-way "sky channel" to its secularized logical conclusion. As early as the 1980s, the French sociologist-philosopher Jean Baudrillard pointed out that the television viewer's symbolic apotheosis as a receiver arrived when TV shows introduced canned laughter. He noted that television had reached a point at which the TV show itself was both the performer and its own audience, "leaving the viewer with nothing but pure amazement."[33]

Even though the Net is sometimes also referred to as "virtual reality," nowadays the television viewer just as frequently experiences his situation as virtual, in the sense of *unreal*. As it stands now, watching television characteristically elicits a feeling that what is being seen must be meant as some kind of absurd parody of what television could be at its worst.

The experience of unreality is enhanced by the glaringly obvious way in which television has become part of the economy. Increasingly, television companies operate on the same pure profit-motive basis as any other business. The essential thing for them is the viewer ratings, because they enable them to sell commercials. Programs have fundamentally become advertisements for the commercials, and viewers are needed only to raise the price of time. An important motivation for traditional media to expand operations onto the Net is the fact that these new technologies offer the chance to gather very detailed information on users, which thus enables the sale of even more precisely targeted advertising. The goal here is to use technology to enhance market-driven audience segmentation.

Since television is connected so closely to capitalism, it

is also, in large part, dominated by the Protestant ethic. This connection illuminates the previously discussed threats to freedom of expression and privacy by contextualizing them as yet another confrontation between the Protestant and the hacker ethic. The commercial character of the media both prevents any focus on commercially uninteresting regions or subjects and leads to violations of privacy.

But one can also argue that if our lives were not so determined by the Protestant work ethic, people would not put up with the current offerings of television. Only when work uses up all energy and people are too tired to enjoy the pursuit of their passions are they ready to be reduced to the passively receptive state suited for television.

The rise of the network society does not in itself give any reason to believe in the general illusion, promulgated in books such as Jeremy Rifkin's *The End of Work*, that the role of work would be automatically diminishing in our life and that our energy thus would be freed for more leisurely pursuits. In fact, in the last couple of decades actual work time has not become shorter but has actually become longer. Any claim of a reduction in working hours can be justified only by a comparison with nineteenth-century industrial society's most extreme twelve-hour workday, but not when it is seen in a more general historical or cultural context.

Furthermore, the mere duration of work time does not provide a comprehensive point of comparison. We must remember that any abbreviation of work time has always been made at the cost of optimizing the remaining work time even further. Shorter working hours do not by any means imply that people are working less. On the contrary: even though working hours have grown shorter than they were in the industrial society's worst-case scenario, they have been optimized to be even more demanding on a person than before. Shorter working hours do not mean a diminution of work or work-centeredness if the same (or even greater) results must be achieved in less time.

In his book *Closing the Iron Cage: The Scientific Management of Work and Leisure,* sociologist Ed Andrew analyzes how the nature of work guided by the Protestant ethic returns us easily to a passive lifestyle in yet another way: "It is not that sociologists of leisure are wrong to think that many workers are incapable of expansive enjoyment off work but rather that they do not take sufficiently seriously the view that incapacity for leisure is a 'spillover effect' of externally managed work."[34] When the individual at work is still treated as a dependent receiver, a trend is encouraged in which leisure is also reduced to passive amusement, with no room for active passions. According to Andrew, only when an active work model has been achieved will active leisure also be realized: only when individuals become self-directed in their work will they be able to become active creators in their leisure time.

The lack of passion in leisure time is truly doubly tragic

when it results from a lack of passion during working hours. In this case, the Friday-centeredness of life is realized in the most absurd way: managed in their work externally, people wait for Friday in order to have more time to watch television and be externally amused. Hackers, on the other hand, use their leisure—Sunday—as an opportunity to realize personal passions other than those that they pursue in their work.

CHAPTER 6

The Spirit of Informationalism

Self-programmable Workers

There is still another part of the hacker nethic to understand: the relation to the network society's networks, other than the media, especially the economic network that affects everyone's lives. Here some computer hackers may feel that the concept of *the hacker ethic* is being extended to concepts beyond those they would normally mean by it. This is absolutely true: these are not typical computer-hacker themes. But from the societal perspective, these themes, defended by only some computer hackers, form an important part of the total challenge of the hacker ethic.

It is useful to start by first characterizing the current dominant reality of these economic networks as they appear to information professionals and to approach the hacker ethic only subsequently. In the typical late-industrial trajectory of a working life (though of course it was never completely realized this way), a person was trained for a job at which they worked for the rest of their productive life, from nine to five. In the information economy, this is no longer the case: instead, the new information professional is, in the words of Castells, "self-programmable" and has "the ability to retrain itself, and adapt to new tasks, new processes, and new sources of information, as technology, demand, and management speed up their rate of change."[1]

In the information age, almost all knowledge becomes outdated quickly, so in order to keep up with the new challenges of their changing projects, the self-programmable need to reprogram their expertise constantly. These challenges of this speedy time are combined with the equally demanding challenges of flexible time. In the new flexible work arrangements—such as telecommuting from home—information professionals must learn to be partly their own managers and to program themselves more efficiently on behalf of the manager.

No wonder that some of them look for help in the self-programming or personal development ("PD") literature. In a time that is shifting from the traditional *personnel* management to *personal* management, it is not surprising that PD books such as Stephen Covey's *Seven Habits of Highly Effective People* and Anthony Robbins's *Awaken the Giant Within* are bestsellers year after year, and that at any given time some new PD book defends a top spot on the bestseller lists. In the information age, there is a need to transfer from the old Taylorist question about physical work—"Could the worker's limbs move in even more optimal trajectories?"—to a more mental one: "Could the person's inner life move in even more optimal trajectories?"[2] As there seems to be something very characteristic about our time in this phenomenon of programming one's self, let's examine its nature in a bit more detail.

The Seven Habits of Personal Development

When we read PD guides, we can find seven key virtues that they teach. Not quite coincidentally, these happen to be the same that were taught by the older Protestant ethic through Franklin, and they can once again be traced back to the monastery. The common starting point for these life methods is *determinacy*, or goal orientation. Individuals are taught to set a well-defined goal and then to direct all their energies toward the achievement of this goal: "Setting goals is the first step," says Robbins,[3] and, to be as exact as possible, the setting of the goal requires a predetermined schedule. Franklin also recommended such planning: "I have always thought that one man of tolerable abilities may work great changes, and accomplish great affairs among mankind, if he first forms a good plan, and, cutting off all amusements or other employment that would divert his attention, makes the execution of that same plan his sole study and business."[4] The PD guides teach one to constantly remind oneself of the goal by, for example, saying it aloud daily and visualizing success in advance.

(In the monastery, this method was called "the remembrance of God." It is striking to note the similarities. Just like the PD gurus, the fourth-century monk Evagrius Ponticus advocated contemplating this desired goal and its opposite by visualizing them: "Imagine the fearful and terrible judgement. Consider the fate kept for sinners. . . . Consider also the good things in store for the righteous. . . . Keep in mind the remembrance of these two realities."[5] The word *vision* itself, before its present PD meaning, referred specifically to the Christian visions of Heaven and Hell. And when PD recommends repeating the goal to oneself every morning, it actually recommends a form of secular prayer.)

According to PD, it is important to remind oneself of the virtues that will help in achieving the goal. One of the most important of these is *optimality*. PD teaches one to make the most focused use of time, so that it always best furthers the work toward the goal. In practice, this means a constant awareness of what use is made of each "now" moment. Robbins exhorts one to remember that "*now* is the time."[6] The main question is, Does what you do at this very moment get you closer to the Goal? If it does not, don't do it; do something else that will.

Franklin taught a similar watchfulness over the "now" moment: "Constant vigilance was to be kept up" and "Be always employ'd in something useful; cut of all unnecessary actions."[7] The PD guides offer the method of contemplating the applicable aphorisms of one's role models and what they would do to get psychic strength for the moment at hand. (In the monastery, this was called the "watch of the heart." The monks were also told to consider whether their actions at any given moment served the highest goal. For example, the sixth-century monk Dorotheus of Gaza exhorted: "Let us pay heed to ourselves and be vigilant, brothers. Who will give us back the present time if we waste it?"[8] In the manner of later PD teachers, Anthony of the Desert recommended in the third century the contemplation of role models so as to be able to act in the present moment according to the supreme goal: "Be mindful of the works of the saints that your souls being put in remembrance of the commandments may be brought into harmony with the zeal of the saints."[9] French classicist Pierre Hadot, who has researched the spiritual exercises of monastic orders, notes that it was exactly to this end that a literary genre consisting of brief biographies of monks was created.[10] Present-day literature about successful CEOs is our hagiography, and collections of their aphorisms are our *apophthegmata,* "sayings of the fathers.")

Other goal-furthering PD virtues are flexibility and stability. Robbins states that the goal should become a "magnificent obsession."[11] In the means used to reach it, however, one must be willing to be *flexible.* Robbins emphasizes that nothing can prevent you from reaching the goal if "you continue to change your approach until you achieve what you want."[12] One must always be ready and humble enough to learn better approaches. Franklin, too, advised one to "perform without fail what you resolve"[13]—whatever flexibility and learning might be necessary on the way. (This was also the attitude of Anthony, who was always willing to learn humbly and to change flexibly in order to get closer to God: "For often he would ask questions, and desired to listen to those who were present, and if any one said anything that was useful he confessed that he was profited.")[14]

Stability means a steady progression toward the goal, which must be kept firmly in view, and setbacks must not be allowed to sway one's emotions in a disruptive manner. From the PD point of view, "negative emotions" such as grief must not interfere. For example, grieving over the loss of something or over some failure does not bring those things back and does not reverse the failure. PD sees negative emotions as a waste of energy that only delays the reaching of the goal.

PD literature teaches a supercharged form of positive thinking to reinforce stability. Robbins, for instance, advises the reader to change negative emotions into positive ones by employing different ways to describe them: *I'm feeling depressed* becomes *I'm feeling calm before action; sad* becomes *sorting my thoughts; I hate* gets transformed into *I prefer; irritated* translates into *stimulated; terrible* reads as *different,* and so on.[15] Again, Franklin urges one to stay calm: "Be not disturbed at trifles, or at accidents common or unavoidable."[16] (Compare this to Cassian, who discourses at length upon the undesirable sin of sadness and the need to replace it with a positive outlook. According to him, sadness may either be "the fault of previous anger" or "spring from the desire of some gain which has not been realized." In either case, it must be put aside, because it does not lead anywhere. Cassian compares the sad soul to "the garment that is moth-eaten [and] has no longer any commercial value or good use to which it can be put.")[17]

Industry is the fifth central virtue in the PD worldview. Striving to reach one's goal, one must admire hard work. Robbins emphasizes how important it is for the individual to be "willing to take massive action."[18] Franklin also lists industry as a virtue. In the opening pages of his *Protestant Ethic and the Spirit of Capitalism,* Weber cites the biblical saying quoted by Franklin's father—"Seest thou a man diligent in his calling, he shall stand before kings"[19]—as an example of the value the Protestant ethic assigns to work. In PD, work is idealized to a degree that sometimes makes it seem like a goal in itself. (This was shared by the monastery, which even counted the opposite of industry, so-called *accedia,* which meant not only laziness but also boredom and restlessness, among the seven deadly sins. This is how Cassian describes its bad effect on monks: "And whenever it begins in any degree to overcome any one, it either makes him stay in his cell idle and lazy . . . or it drives him out from thence and makes him restless and a wanderer.")[20]

The value of *money,* stressed by Franklin in his Protestant ethic, also figures prominently in PD. Robbins subtitles his book *How to Take Immediate Control of Your Mental, Emotional, Physical & Financial Destiny!* In the models provided for goal selection in PD guidebooks, money serves as the chosen example for a goal. In Robbins's goal-setting form, money is a built-in objective:

> Do you want to earn:
> $50,000 a year?
> $100,000 a year?
> $500,000 a year?
> $1 million a year?
> $10 million a year?
> So much that you can't possibly count it?[21]

(The monastic life's connection to economy is more complex than in the case of the other virtues. The goal of the monasteries was not to make money, but it is not just an accident that the word *economy,* based on the Greek word *oikonomia,* is used in theological parlance in reference to the doctrine of salvation. In both capitalism and the monastery, life is subordinated to the striving for "salvation" or "heaven"—that is, to the economic end.)

In the PD world, nothing is left to chance in the realization of the goal and its attendant virtues; everything must be accounted for. Therefore, *result accountability* is the seventh important virtue. Readers of Robbins's book explicitly put their goals down in writing and go on to account for their progress toward them. This is how Robbins suggests one should document the evolution of one's emotions:

1. Write down all the emotions that you experience in an average week.
2. List the events or situations you use to trigger these emotions.
3. Come up with an antidote for each negative emotion, and employ one of the appropriate tools for responding to the Action Signal.[22]

Once again, Franklin's shade hovers behind this. In his *Autobiography*, Franklin tells us how he wrote down his goals: "I form'd written resolutions, which still remain in my journal book."[23] He also tells us how he realized that it was not enough just to write down the goals and virtues, but that for their realization "daily examination would be necessary."[24] In his *Autobiography*, he describes the spiritual bookkeeping he devised to this end:

> I made a little book, in which I allotted a page for each of the virtues [which included among others the aforementioned virtues of resolution and tranquillity]. I rul'd each page with red ink, so as to have seven columns, one for each day of the week, marking each column with a letter for the day. I cross'd these

> columns with thirteen red lines, marking the beginning of each line with the first letter of one of the virtues, on which line, and in its proper column, I might mark, by a little black spot, every fault I found upon examination to have been committed respecting that virtue upon that day.[25]

(Compare this to how the monks were also taught to systematically observe their progress. Dorotheus writes:

> We ought not only to examine ourselves every day but also every season, every month, and every week, and ask ourselves: "What stage am I at now with regards to the passion by which I was overcome last week?" Similarly, every year: "Last year I was overcome by such and such a passion; how about now?" The [Church] Fathers have told us how useful it is for each of us to purify himself in turn, by examining, every evening, how we have spent the day, and every morning, how we have spent the night.[26] We can consider modern result accounting as a form of secular confessing, an office confessional.)

Finally, it is important to note that the emphasis on being methodical links the monastery and PD in one more important respect: in both cases, the method offers the promise of an experience of clarity and certainty in the world. Looked at purely from this point of view, it does not actually matter which method a person believes in firmly.

Salvation has been realized both in the monastery and through PD. It seems that there is increasing demand for such clarity and certainty in an era of ever more complex networking at ever greater speeds. It seems that the more complicated and speedy our exterior development becomes, the greater the demand for interior simplicity grows.

Through PD, the complicated and speedy world is managed by teaching people to pursue ever more specific goals. If individuals are to make their marks in a world of global competition, they must "localize" their goals ever more pointedly. They must each concentrate on one fixed point and exclude most of the rest of the world. Speed is managed by focusing on the moment at hand. Life becomes manageable when it is reduced to *one* goal and *one* moment at a time. The question, then, is simply, Am I living right now according to my highest goal? PD takes this even a step further by also giving fixed answers for every situation (flexibility, stability, etc.).

This religious tone of PD makes it clear that even though the PD method is aimed at the achievement of the goal at hand, psychologically it is not just instrumental. Spiritually, life becomes easier in the network society if one has recourse to some clear-cut method in whose powers of salvation one can believe unconditionally, and this is the reason both PD teachings and fundamentalism have become ever more attractive in the network society.

The Spirit of Informationalism

Some may have wondered why we should bother to analyze PD in the context of the network society. The reason is that this examination can throw some indirect light on the central issue of the logic of economic networks that Castells raises in *The Information Age*. He asks what the " 'ethical foundation of the network enterprise,' this 'spirit of informationalism' " is and goes on to specify: "What glues together these networks? Are they purely instrumental, accidental alliances? It may be so for particular networks, but the networking form of organization must have a cultural dimension of its own." The same question can be asked even more generally about the spirit of the network society, which is built on informationalism, the new information-technology paradigm. Castells himself leaves this central question unanswered by saying only that the spirit of informationalism is "a culture of ephemeral," which is unfortunately the same as saying that it does not have any collective or permanent values.[27]

Of course, we must be aware that it is not at all easy to describe the dominant spirit of a time, and it is particularly difficult to do so with the values of the network society, which functions in diverse cultures with diverse values, in an era in which these values are additionally subject to rapid transformation everywhere. So it easily appears at first that the network society is a totally valueless society: network enterprises are willing to adapt their products to the values of any culture at all (different versions of a product are marketed in different countries by appealing to local cultural values) and are even willing to commodify some of those cultural values themselves if a sufficient market can be found (such as that for exotic commodities). At the same time, cultures are in the process of abandoning any traditional values that impede the activity of network enterprises in their sphere, in order not to be left out of the global information economy.

However, while considering the spirit that governs network enterprises, one does well to remember that when Weber used the term *the spirit of capitalism* or *the Protestant ethic*, he did not refer to a culture that had evolved everywhere in exactly the same way. It was not his intention to claim that all cultures governed by the spirit of capitalism and the Protestant ethic shared all of the same values. Second, the values that he found gluing development together were very different from the old ethical values: work and money.

With these clarifications, it is possible to characterize the values that guide the network enterprises and even the network society more generally, even though the network society may contain many other values within its varied cultural manifestations. There are reasons to say that the network enterprise is held together by the same seven values PD teaches in an exaggerated form: goal orientation, optimality, flexibility, stability, industry, economy, and result accountability. And these *are* values in the traditional philosophical sense: the overriding goals guiding action—even though they do not resemble the old ethical values.

To an ever increasing degree, this list also describes the values of states—the new form of which Castells calls "the network state"[28]—and so they can be seen as embodying the dominant spirit of the whole network society. The spread of this spirit from enterprises to states is not surprising, since the reason traditional nation-states have delegated power to networks of states such as those formed by the European Union, the North American Free Trade Agreement, and the Asia-Pacific Economic Cooperation, has to a great extent been to prosper better in the information economy. The actions of states are increasingly governed by economic goals.

These seven values can be said to have an internal hierarchy: *money* is the highest value or goal of the network society's governing spirit, and the other values support the realization of that goal. Among the other values, *work* still has special status: states, in particular, still champion it as an independent goal, but even on that level it is slowly becoming more and more clearly subordinated to money. Just like the network enterprise as a form, optimality, flexibility, stability, determinacy, and result accountability can be seen as consequences of capitalism's adaptation to making money in a new technological situation.

Here, Robbins's advice to the individual provides a good expression for this way of thinking about values:

"What do my values need to be in order to achieve the destiny [money] I desire and deserve? . . . See which values you might get rid of and which values you might add in order to create the quality of life you truly want." And: "What benefit do I get by having this value in this position on my hierarchy?" In this view, values are purely instrumental for amassing money—something Weber already recognized in Franklin's value system.[29]

Thus, while the information economy introduces new values to the ones of the spirit of the old capitalism, these are essentially designed to guarantee the continuity of the old goal to make money. As a goal, money—an instrumental value—is a peculiar one: when society's vision consists of the maximization of money, the realization of this vision does not require any real changes in the world. This is connected to the value of *flexibility.* Commercial enterprises and states do not talk about changing the world; they have progressed to a flexible strategic mode of thinking that is designed to safeguard the continued success of moneymaking in any possible world. If one approach does not work, the enterprise and the state are ready to change, and other ways of thinking become labeled naïve idealism.

In the fast-paced competition of the information economy, modes of operation have to be dynamic. This leads to the organization of operations into projects, and these for their part require ever greater *goal orientation* and *result accountability.* This goes for both the main projects the whole enterprise has embarked on and for individual workers' engagement in their partial projects. Projects must have clear-cut goals and schedules, and their progress must be followed systematically. This becomes increasingly important when information professionals have more freedom to choose the times and locales of their work: goals and deadlines become essential determinants of the working relationship. These modes are also, gradually, becoming more prevalent in the ways states operate.

Optimality is important to network enterprises. Self-programming enters the picture again: network enterprises optimize their functions in the way that computer and network operations are optimized. The new business thinking of dot-com capitalists can actually be seen as reprogramming the process. The dot-coms examine the stages of business processes as if they were lines of programming code: unnecessary ones (e.g., in distribution, wholesalers and retailers) are eliminated, and slow routines are rewritten from an entirely new point of view to make them work faster.

The organization of employment relationships is also optimized, as if it were a question of improving a computer network. Employment relationships are seen as a situationally fluctuating network of resources. To their own core skills, enterprises connect other skills or disconnect them as necessary. The optimization of both process and organization in this manner has become possible since governments have underwritten the idea of a flexible labor force.

Stability completes the list of values defining the network society's dominant spirit. At the governmental level, this ideal manifests itself in the way politicians have replaced their former usage of words such as *justice* and *peace* with the new term *stability.* The EU wants stability in Europe's development (e.g., for Yugoslavia there is the Stability Pact for South Eastern Europe).[30] The United States wants to stabilize conditions in various parts of the world, and the same stability is seen as desirable in the development of Asia. Internally, governments worry that the divide between the successful and unsuccessful increases "social instability." This, again, is undesirable mainly because it presents a threat to the realization of the monetary goal—instability, as we know, is not welcomed by companies, which take fright at the thought of volatility.

Against this background, it can now be understood that the value system of PD works so well for successful workers in network enterprises because they are, in fact, an application of the enterprises' own values to the individual's own life. Within PD, a person treats his or her life as if it were a network enterprise, asking, What is my vision? What is my strategy for its realization? Life becomes a project with quarterly progress reports.

In the end, the ideals of a network enterprise or person and those of a computer or network are actually the same:

the ability to function flexibly in a way optimal for each project goal, while maintaining stability at high speed. It is this fact that gives us a reason to speak of the spirit of informationalism, which refers to the new technological basis of our society, especially the networks of computers. Both the network enterprise or state and the people practicing PD apply the informationalist metaphors of the computer and network to themselves.

This is ultimately what makes PD and the dominant spirit of the network society questionable: the problem is not that these principles could not lead to the achievement of goals; the problem is its definition of what it is to be human. In PD and the spirit of the network society, the logic of a computer network–based society is applied to humans and their relationships. The human being is treated like a computer, with mental routines that can always be reprogrammed in a better way. It would be possible to translate the entire body of PD teaching into a short computer program that human beings are supposed to run. Robbins talks explicitly about the human being as a "mental computer."[31] The idea of the human being as a computer is extended in PD to human relationships by treating them as computer networks. Robbins writes: "I've found that, for me, the greatest resource is a relationship because it opens the doors to every resource I need."[32] Thus, the previously discussed values active in an individual's own actions are also applied to her or his human relationships: one should connect with people who are useful to one's goal and disconnect from those who are useless or even detrimental ("bad company").

The Ethics of the Network

Among the seven values we have discussed, stability is the closest to the old ethical values. Nevertheless, it differs from them in ways that aptly demonstrate the difficult time real ethics have in the network era. A network is stable when it does not crash and bring the activities pursued within it to a halt. Similarly, our new ideal is a society that is stable in that it does not interfere with the financial market's functioning in the global computer network.

Let's see in more detail what the application of the network metaphor to people and society means for ethics. Network logic requires constant optimization by connecting and disconnecting resources as needed, the only limitation being the need to keep the network stable. In practice, it is difficult to realize this without at the same time replacing ethics with a philosophy of survival. Business enterprises optimize their networks in order to survive in economic competition, and the ones that cannot keep up are left outside the networks. The ironic culmination of this survival logic derives from the fact that the more networks end up incorporating only the information elite, the more that elite itself must also be concerned about survival. The information professional can be reminded of this survival aspect when some excluded person unexpectedly threatens him with violence in the street or in front of his own home in broad daylight. For a moment, the outcast from the network society has power: the professional finds his information-processing skills seriously challenged as he searches for the right words to extricate himself from this physically threatening situation. Facile solutions to this problem rely on a reinforcement of "stabilizing" factors: more police officers are hired, and the high-level elite resorts to its own bodyguards. At a global level, the most developed countries "stabilize" the wars among the outcasts depending on how important each conflict is to the global economy.

To connect this *logic of exclusive networking,* some hackers defend the goal of inclusive networking. The case in point is the hacker institution at the heart of the Net's development, the Internet Society. Its ethic is expressed by the principle "No discrimination in use of the Internet on the basis of race, color, gender, language, religion, political or other opinion, national or social origin, property, birth or other status."[33] The Internet Society supports the diffusion of the Net and the teaching of network skills to all who have been left out of the development of enterprises and governments. This is an enormous task. At the moment of this writing, only about 5 percent of the world's people have access to the Net (of which about half are in North America; Africa and the Middle East together have fewer users than there are people in the Bay Area), and half of the world's adult population has never even used a telephone.[34] Thus, in practice, hacker endeavors have not made much of a difference as yet, but NetDay, kind of a new Labor Day celebrated annually by some hackers to remind us of this task, is an important symbol of the ideal of caring for everyone as an end in itself and not just for stability.[35] Of course, mere technical networking will not be enough to make a society just, but it is a necessary prerequisite for achieving fairness at the level of the economic networks, which is the level of the worker's relation to the company.

The Ethics of the Computer

The application of the computer metaphor to people and society makes real ethics also very difficult. The optimization of human beings and enterprises in terms of the computer leads to *the logic of speed,* and this tends to make our lives survival-based in another way. At high speeds, the societal goal becomes the same as the one pursued by race-car drivers: to keep the vehicle stable so as to prevent it from running off the track. Here, the ideal of stability threatens to replace ethics once again.

One might say that there is an "ethics barrier," a speed above which ethics can no longer exist. After that point, the only remaining goal is to survive the immediate moment. But only those who do not have to focus purely on the "now" to guarantee their own survival are able to care for others. Ethicality requires unhurried thinking.

Ethicality also requires a longer temporal perspective: responsibility for the future consequences of prevailing developments and the ability to imagine the world as becoming different from the way it is now. In regard to this second profound problem of our era, hackers are again only able to provide a more or less symbolic example of a different, more caring relationship with time. For example, Danny Hillis has noted that humanity is moving at such developmental speed that it is unable to see anything except what is already here or, at best, what will be here in a couple of years, thanks to the already prevailing speed. He wrote in 1993: "When I was a child, people used to talk about what would happen by the year 2000. Now, thirty years later, they still talk about what will happen by the year 2000. The future has been shrinking by one year per year for my entire life."[51]

To counter this, hackers have traditionally reserved time for thought experiments in regard to even the distant future. We know that computer hackers have always felt at home in future research fields and that many of them are great fans of science fiction. Thus, it comes as no surprise that a group of hackers joined Hillis to start the Long Now Foundation, the motive of which is to shake up our time perspective. The foundation's main project consists of building a clock symbolizing and encouraging thought in a long perspective. Hillis wrote: "I want to build a clock that ticks once a year. The century hand advances once every 100 years, and the cuckoo comes out on the millennium. I want the cuckoo to come out every millennium for the next 10,000 years."[37] Brian Eno, father of ambient music and another founding member, gave the clock its name: the Clock of the Long Now. Other figures behind the clock include Mitch Kapor and Stewart Brand, who, as we have seen, were also the founders of the Electronic Frontier Foundation.

The various designs proposed for the actual timepiece have ranged from a gigantic clock mechanism in the California desert to Peter Gabriel's suggestion of a garden in which short-lived flowers indicate the passing of seasons and giant redwoods the passing of years. Recently, the foundation has finally decided to acquire a site for the clock adjoining the Great Basin National Park in Nevada.

The main thing about the clock is, of course, not its mechanism but its ability to attune us, symbolically, to a different sense of time. It is meant to be an ethical symbol, similar to the first images of our blue globe published by NASA in 1971. Those images made us see the earth both as a whole and as a fragile little planet amid the enormity of outer space, which is why environmental groups chose such images as their symbols. In the Clock of the Long Now, technology is removed from the network society's dominant model of time and made to serve a rhythm that gives caring a chance. It leads us from the ideal of retaining stability at high speed to genuine ethical being.

Caring

In addition to the annual NetDay and the Clock of the Long Now, there is a third important hacker expression of caring, opposed to our time's survival tendency. That is the direct caring about those who are on the edge of survival. Some hackers have used the resources they have acquired through capitalism to support those who must literally fight for their survival. Although here, too, hackers' influence has been very limited, they have set an exemplary alternative answer to the question, Why would you want to have a lot of money? They do not take it as self-evident that the answer is to want something for oneself, to buy one's way into being part of the establishment; instead, their answer is that people can direct resources from the egoistic economy toward those who are exploited by it. For example, Mitch Kapor supports a global environmental-health program for eliminating health problems caused by corporate practices.[38] Sandy Lerner, who left Cisco Systems with Leo Bosack in 1990 with $170 million in stock, has used that money to start a foundation that fights the cruel treatment of animals.[39]

The logic of the network and the computer alienate us from direct caring, which is the beginning of all ethical behavior. We need more of the kind of thinking about the peculiar challenges of caring in the information age that some hackers represent. We will do well not to expect these thoughts to come from corporations or governments. Historically, such entities have not been sources of new ethical thinking; instead, fundamental changes have been initiated by some individuals who care.

CONCLUSION

CHAPTER 7

Rest

The Seven Values of the Hacker Ethic

We have seen that the seven dominant values of the network society and Protestant ethic are money, work, optimality, flexibility, stability, determinacy, and result accountability. Now we can summarize the seven values of the hacker ethic that have had a significant role in the formation of our new society and that represent a challenging alternative spirit of informationalism.

Again, it is important to remember that only few computer hackers share them all, but they must be seen collectively because of their societal and logical relation to one another.

Each chapter up to now has concentrated on one of these values. The first guiding value in hacker life is *passion,* that is, some intrinsically interesting pursuit that energizes the hacker and contains joy in its realization. In chapter 2 we discussed *freedom.* Hackers do not organize their lives in terms of a routinized and continuously optimized workday but in terms of a dynamic flow between creative work and life's other passions, within which rhythm there is also room for play. The hacker *work ethic* consists of melding passion with freedom. This part of the hacker ethic has been the most widely influential.

In the hacker *money ethic,* discussed in chapters 3 and 4, the striking element is that many hackers still follow the original hackerism in that they do not see money as a value in itself but motivate their activity with the goals of *social worth* and *openness.* These hackers want to realize their passion together with others, and they want to create something valuable to the community and be recognized for that by their peers. And they allow the results of their creativity to be used, developed, and tested by anyone so that everyone can learn from one another. Even though much of the technological development of our information age has been done within traditional capitalism and governmental projects, a significant part of it—including the symbols of our time, the Net and the personal computer—would not exist without hackers who just gave their creations to others.

As we've seen, a third crucial aspect of the hacker ethic is hackers' attitude toward networks, or their *nethic,* which is defined by the values of *activity* and *caring.* Activity in this context involves complete freedom of expression in action, privacy to protect the creation of an individual lifestyle, and a rejection of passive receptiveness in favor of active pursuit of one's passion. *Caring* here means concern for others as an end in itself and a desire to rid the network society of the survival mentality that so easily results from its logic. This includes the goal of getting everybody to participate in the network and to benefit from it, to feel responsible for longer-term consequences of the network society, and to directly help those who have been left on the margins of survival. These are still very open challenges, and it remains to be seen if hackers can have an influence here on the same scale as they have had on the other two levels.

A hacker who lives according to the hacker ethic on all three of these levels—work, money, nethic—gains the community's highest respect. This hacker becomes a true hero when she or he manages to honor the seventh and final value. It has appeared in this book all along, and now, in the seventh chapter, it can be explicated: it is creativity—that is, the imaginative use of one's own abilities, the surprising continuous surpassing of oneself, and the giving to the world of a genuinely valuable new contribution.

In his manifesto "Deus Ex Machina, or The True Computerist," the Homebrew Computer Club's Tom Pittman expressed the importance of creativity in his description of the feeling that accompanies true hacking: "In that instant, I as a Christian thought I could feel something of the satisfaction that God must have felt when he created the world."[1]

In its attitude toward creativity, the hacker ethic differs

once again from the Protestant and pre-Protestant versions. Pittman's grandiose simile enables us to finish the playful arc of this book by taking the freedom of placing these three ethics inside the same metaphorical landscape, that of Genesis, with which the discussion of the hacker ethic began in chapter 1. Almost needless to say, this approach will go further than most computer hackers would, but in the concluding chapter of a book that deals with the broad, basic questions of our philosophy of life, such a mythical dimension is only appropriate.

The Protestant Genesis

Genesis is a rich myth, and it puts in an appearance whenever the most profound questions about what it means to be human arise. In the first chapter, we saw how important a mirror it has been, historically, for the description of our work ethic. Similarly, our concepts of creation and creativity down the ages are well reflected in it.

In the pre-Protestant time, Augustine was bothered by the question of why God created the world precisely when he did.[2] In the eighteenth century, the Protestant Dr. Lightfoot wanted to calculate the exact moment of the event. Using the Bible, he arrived at the conclusion that the world was created on Friday, October 23, 4004 B.C., at 9 A.M. Of course, it was very fitting to the Protestant ethic to come up with Friday as the creation day: the world was created at the beginning of a workday because it was designed for work.

Regarding work as an intrinsic value, the Protestant ethic implies that the leisured state humanity lost in the fall really wasn't a loss. Milton asks in his seventeenth-century Protestant epic *Paradise Lost,* Why would God have planted a forbidden tree in the middle of Paradise if humans had not been meant to eat its fruit?[3] The Protestant ethic's answer is that humans were indeed meant to eat of the fruit: to work by the sweat of their brows was their true and intended lot.

In the evolution of the Protestant ethic, Paradise can even be seen merely as a lesson to show Adam and Eve how undesirable idleness really is. When a human being is idle, she or he finds a surrogate activity—eating—and then accuses another person of its damaging consequences. A life in which one is alone responsible for its contents is the most difficult one of all. In the Protestant ethic, a job seeker is not just a seeker of a job but a seeker of a solution to the problem of his or her life. Work offers an answer: the meaning of life is welding or bookkeeping or being a CEO or what have you. With the help of work, a person's identity becomes defined practically. Working, no one has to wake up to worry about how to live each day.

In a world governed by the Protestant ethic, we work because we do not know what else to do with our lives—just as we live because we do not know what else we would do. We work to live—a life consisting of work. In other words, we work in order to work, and live in order to live. One must hope that the preacher Wilhelm Schneider was not right when he claimed that even in the heavenly life to come we need work so that eternity might not feel so long![4]

Creativity does not feature prominently in the Protestant ethic, the typical creations of which are the government agency and the monasterylike business enterprise. Neither one of them encourages the individual to engage in creative activity.

The anticreativity of these institutions can be suggested by a thought experiment: how would they have gone about the creation of the world? The beginning of a government agency's Genesis, involving endless meetings and proposals before action, would look something like this:

MINUTES OF THE INAUGURAL MEETING OF THE COMMITTEE DEALING WITH THE SUBJECT OF THE CREATION OF THE WORLD

Time: 23 Oct 5004 B.C. 9 A.M.
Place: Heaven, Sphere 9
Present:
God (Chair)
Archangel Michael
Archangel Raphael
Archangel Gabriel (Secretary)
Absent:
Lucifer

1. Opening of the meeting
   God opened the meeting and welcomed the participants at 9:00 A.M.

2. Approval of the proposed agenda
   The proposed agenda was approved as the agenda for this meeting.
3. The creation of the world
   A lively debate ensued on the Chair's idea to create a world. It was decided to form a committee dealing with the subject of creation, entrusted with the task to prepare a world-creation strategy developed from this initial idea. The strategy is to narrow its focus on the world and on how everything ought to be.
4. Other matters
   It was decided to switch from muffins to doughnuts with our coffee, and to invite bids.
5. Next meeting
   The next meeting will be at the end of the world.
6. End of meeting
   The Chair closed the meeting at 12:00 noon.

Signed, Archangel Gabriel, Secretary

STRATEGY FOR THE CREATION
OF THE WORLD—SYNOPSIS

The reader is now holding the synopsis of the strategy for the creation of the world. The more extensive justifications for this strategy have been published separately in a series of reports by God's Research Foundation, which include the expert opinions of angels solicited in the process of formulating this strategy.

The strategy starts out from the recognition that the world must be created on the basis of content rather than technology. In the long run, a mere technical infrastructure, such as earth, light, and the overarching lid, are not sufficient. Skill in content creation is needed. Therefore, life should be developed as the world's content by means of six spearhead projects.

VISION OF THE WORLD

There is life in the world, whose task is to bring life into the world.

SUGGESTED ACTIONS TO BE TAKEN

Creation will be advanced by means of the following six high-profile spearhead projects:

1. Creation of heaven and earth
2. Creation of light
3. Creation of the heavens
4. Creation of plants
5. Creation of animals
6. Creation of human beings

The strategy committee's action plan is that working groups be established in the next phase for each one of these spearhead projects.

In the commercial enterprise version, the Bible would begin with a contract, in which creation would be mentioned only as an introduction to agreements on who gets what:

CONTRACT

The creator of the world (henceforth "God") and the parties granted use rights to the world (henceforth "human beings") have agreed this day 27 February 2347 B.C., after the flood, the following:

PURPOSE OF CONTRACT

1. The human beings promise to repent their sins and live more righteously from now on. Repentance and penitence are to be completed by the agreed-upon deadline: the span of each human's lifetime.
2. God grants the human beings grace, consisting of the following two elements:
   —refraining from further floods
   —eternal life

God will grant this grace in two installments. The first installment, i.e., the restraint from further floods, will be granted on signature of contract. The second installment, i.e., eternal life, will be granted when human beings' performance has been approved at the end of the world.

RIGHTS

3. The distribution and use rights of the grants mentioned in point 2, above, i.e., forgiveness and eternal life, will remain entirely with God. All rights to the product names World and Eternal Life are likewise the sole property of God.
4. Protection of competitive advantage: human beings will not enter into any agreements concerning objectives similar to those expressed in this contract with any parties in competition with God.

SANCTIONS

5. Should human beings prove unable to fulfill the duties defined in this contract, God reserves the right to torture them as much as he wants in all the ways he may invent throughout eternity. No rights involving sanctions are vested in the human beings.

RESOLUTION OF CONTRACTUAL CONFLICT

6. Any conflicts arising out of this contract will be resolved in Helsinki Circuit Court.

27.2.2347 B.C.

Signed:

____________________	____________________
God	For the human beings Noah

Witnessed by:

____________________ ____________________

Shem Ham

In the hacker model, the individual simply starts creating, without any bureaucratic formalities, and passes her or his creation on to others directly without any complicated legalese.

The Pre-Protestant Genesis

The pre-Protestant view of creation also differs from the Protestant ethic. According to the pre-Protestant Church Fathers, God did not act on Friday; rather, the paradisiacal world in which human beings were not meant to do anything was created, appropriately enough, on Sunday. Sunday is also the day on which Christ rises to his rest in Heaven. In his *Apology* for Christianity in the second century, one of the Church Fathers, Justin Martyr, praises Sunday for both of these reasons:

> Sunday is the day on which we all hold our common assembly, because it is the first day on which God, having wrought a change in the darkness and matter, made the world; and Jesus Christ our Saviour on the same day rose from the dead.

The Protestant ethic celebrates Friday; the pre-Protestant one sanctifies Sunday. This evaluative difference is also expressed in the way that Sunday was regarded as the *first* day of the week in the pre-Protestant era, whereas it is now commonly seen as the *last* day of the week.

While the Protestant ethic is work-centered, one might see the pre-Protestant ethic, then, as leisure-centered. This leisure-centeredness does not encourage creativity any more than work-centeredness, however, as it is defined negatively, as not-work, rather than in terms of some positive use. The effect of this attitude can be seen in the relative absence of creativity during the first millennium and a half after Christ, most remarkably in the field of science. Quite typically, the question that most engaged pre-Protestant Church Fathers, in the wake of Augustine, was, *Why* did God create the world? From a pre-Protestant viewpoint, this was a genuine problem: logically, the pre-Protestant God would have valued leisure so highly that he would not have bothered to create anything.

Beyond Friday and Sunday

We have throughout this book used the metaphorical expression that hackers defend Sunday against Friday, although this statement has always been qualified. An examination of the Protestant and pre-Protestant ethics in regard to creation clarifies the importance of these qualifications and shows the important ways in which, in the end, the hacker ethic differs from *both* the spirit of Friday and that of Sunday.

From the hacker viewpoint, leisure-centeredness may be just as undesirable as work-centeredness. Hackers want to do something significant; they want to create. While they avoid work that does not give rise to an opportunity for creativity, they also consider leisure as such insufficient as an ideal state. A Sunday spent in apathetic leisure can be as insufferable as a Friday. The very idea of Heaven as this kind of eternal Sunday has made many atheists agree with Machiavelli that they would rather go to Hell (often thinking of Dante's forecourt of the Inferno, in which the greatest philosophers and scientists of antiquity are still allowed to pursue their creative investigations).[5]

Hackers do not feel that leisure time is automatically any more meaningful than work time. The desirability of both depends on how they are realized. From the point of view of a meaningful life, the entire work/leisure duality must be abandoned. As long as we are living our work or our leisure, we are not even truly living. Meaning cannot be found in work or leisure but has to arise out of the nature of the activity itself. Out of passion. Social value. Creativity.

Pittman's approach to Genesis describes this brilliantly. Based on its tone, we can say that the hackers' answer to Augustine's question is that God, as a perfect being, did not *need* to do anything at all, but he *wanted* to create. In the hacker attitude, creativity is an intrinsic value. For a description of its psychology, one can read the beginning of Genesis not as a description of the creation of the world but, less grandiosely, as the experience of creative action:

> And the earth was without form, and void; and darkness was upon the face of the deep. And the spirit of God moved upon the face of the waters. And God said, Let there be light: and there was light. And God saw the light, that it was good.[6]

In Genesis, when at the moment of the arrival of the creative idea darkness changes into light, God cries out like any creative artist: "Yes! There it is!" He is not just anybody: he is *He*. He is proud for a moment: "Well, I seem to be pretty good at making these."

Genesis can be seen as a tale of the kind of activity that occurs on creativity's own terms. In it, talents are used imaginatively. It reflects the joy one feels when one surprises and surpasses oneself. Every day, God comes up with an even more extraordinary idea: how about making some bipedal hairless creatures. . . . And he gets so enthusiastic about the creation of a world for others that he is even ready to stay awake for six nights in a row, getting some rest only on the seventh day.

Because of its emphasis on creativity, the hacker ethic must ultimately be considered distinct from both the Protestant and the pre-Protestant ethics. According to the hacker ethic, the meaning of life is not Friday, but it is not Sunday, either. Hackers locate themselves between the Friday and Sunday cultures and thus represent a genuinely new spirit. We have only just begun to understand its significance.

EPILOGUE

Informationalism and the Network Society

MANUEL CASTELLS

Technology is a fundamental dimension of social change. Societies evolve and transform themselves through a complex interaction of cultural, economic, political, and technological factors. So technology has to be understood within this multidimensional matrix. Yet technology has its own dynamics. The kind of technology that develops and diffuses in a given society decisively shapes its material structure. Technological systems evolve gradually until a major qualitative change occurs: a technological revolution, ushering in a new technological paradigm. The notion of paradigm was proposed by the leading historian of science Thomas Kuhn to explain the transformation of knowledge by scientific revolutions. A paradigm is a conceptual pattern that sets the standards for performance. It integrates discoveries into a coherent system of relationships characterized by its synergy—that is, by the added value of the system vis-à-vis its individual components. A technological paradigm organizes the available range of technologies around a nucleus that enhances the performance of each one of them. By *technology,* it is usually understood the use of scientific knowledge to set procedures for performance in a reproducible form.

Thus, the industrial revolution constituted industrialism, a paradigm characterized by the capacity to generate and distribute energy by human-made artifacts, without depending on the natural environment. Because energy is a primary resource for all kinds of activities, humankind was able, by transforming energy generation, to increase dramatically its power over nature and over the conditions of its own existence. Around the nucleus of a technological revolution cluster and converge technologies in various fields. The revolution in the technology of energy (first with steam power, then later with electricity) laid down the foundations of industrialism. Associated revolutions in mechanical engineering, metallurgy, chemistry, biology, medicine, transportation, and a wide variety of other technological fields came together in the constitution of the new technological paradigm.

This technological infrastructure made possible the emergence of new forms of production, consumption, and social organization that together formed the industrial society. Central features of the industrial society were the industrial factory, the large corporation, the rationalized bureaucracy, the gradual phasing out of agricultural labor, the process of large-scale urbanization, the formation of centralized systems for delivery of public services, the rise of mass-media communication, the construction of national and international transportation systems, and the development of weapons of mass destruction. Industrialism appeared in a variety of cultural and institutional expressions. Industrial capitalism and industrial statism were antagonistic forms of social organization, yet they shared fundamental similarities in their material foundations. History, culture, institutions, and evolving patterns of political domination created a diverse array of industrial societies, as different as Japan and the United States, Sweden and Spain. Yet these were historical variations of a common sociotechnological species: industrialism.

This analogy may help explain the meaning and importance of informationalism as a technological paradigm, which is currently replacing industrialism as the dominant matrix of twenty-first-century societies. To be sure, industrialism does not disappear in one day or in a few years. The process of historical transition proceeds by absorption of preceding social forms by the new, emerging ones, so that real societies are considerably messier than the ideal types we construct for heuristic purposes. How do we know a given paradigm (e.g., informationalism) is dominant vis-à-vis others (e.g., industrialism)? It's simple: because of its superior performance in the accumulation of wealth and power. Historical transitions are shaped by the world of winners. This should imply no value judgment. We do not really know if producing more or more efficiently embodies superior value in terms of humanity. The idea of progress is an ideology. How good, bad, or indifferent a new paradigm is depends on whose perspective, on whose values, on whose standards. But we know it is dominant because when implemented, it erases competition by elimination. In this sense, informationalism is the dominant paradigm of our societies, replacing and subsuming industrialism. But what is it?

Informationalism is a technological paradigm. It refers to technology, not to social organization and not to institutions. Informationalism provides the basis for a certain type of social structure that I name the network society. Without informationalism, the network society could not exist, yet this new social structure is not produced by informationalism but by a broader pattern of social evolution. I will elaborate below on the structure, genesis, and historical diversity of the network society. But let me first focus on its material infrastructure: informationalism as a technological paradigm.

What is characteristic of informationalism is not the central role of knowledge and information in the generation of wealth, power, and meaning. Knowledge and information have been central in many, if not all, historically known societies. There were certainly different forms of knowledge in many instances, but knowledge, including scientific knowledge, is always historically relative. What is accepted as truth today may be cataloged as error tomorrow. To be sure, in the last two centuries there has been a closer interaction than in the past between science, technology, wealth, power, and communication. But the Roman Empire cannot be understood without the technology of engineering of vast public works and communication patterns, without the logical codification of government and economic activities in the Roman Law, and without the processing of information and communication made possible by a developed Latin language. Throughout history, knowledge and information, and their technological underpinnings, have been closely associated with political/military domination, economic prosperity, and cultural hegemony. So, in a sense, all economies are knowledge-based economies and all societies are, at their core, information societies.

What is distinctive in our historical period is a new technological paradigm ushered in by the information-technology revolution, centered around a cluster of information technologies. What is new is the technology of information processing and the impact of this technology on the generation and application of knowledge. This is why I do not use the notions of knowledge economy or information society but the concept of informationalism: a technological paradigm based on the augmentation of the human capacity in information processing around the twin revolutions in microelectronics and genetic engineering. However, what is revolutionary in these technologies vis-à-vis previous information-technology revolutions in history, such as the invention of the printing press? Printing was indeed a major technological discovery, with considerable consequences in all domains of society—although it induced much greater changes in the European context in the early modern age than in the Chinese context, where it was invented much earlier. But the new information technologies of our time have an even higher historical relevance because they ushered in a new technological paradigm on the basis of three major, distinctive features:

1. their self-expanding processing capacity in terms of volume, complexity, and speed,
2. their recombining ability, and
3. their distributional flexibility.

I will now elaborate on these features, which constitute the essence of the new, informational paradigm. I will do it separately for the two fundamental technological fields—microelectronics and genetic engineering—before considering their interaction.

The microelectronics-based revolution includes the microchip, computers, telecommunications, and their networking. Software development is the critical technology to operate the whole system, but integrated circuits hold the processing power in their design. These technologies allow for an extraordinary increase in the capacity to process information, not only in the volume of information but in the complexity of the operations and in the speed of the processing. However, how much is "much more" compared with previous information-processing technologies? How do we know that there is a revolution characterized by an unprecedented leap forward in processing capacity?

A first layer of the answer is purely empirical. Take any of the available measures of information processing, in terms of bits, feedback loops, and speed, and the last thirty years have seen a sustained exponential increase in processing power, coupled with an equally dramatic decrease in cost per operation. But I venture the hypothesis that there is something else, not only quantitative but qualitative: the capacity of these technologies to self-expand their processing power because of the feedback on technological development of the knowledge generated on the basis of the technology. This is a risky hypothesis, because processing power may find physical limits for further integration of circuits in the microchip. However, until now, every doomsday prediction in this domain has been belied by new manufacturing breakthroughs. Ongoing research on new materials (including biological materials, and chemically based information processing on their DNA) may extend extraordinarily the level of integration. Parallel processing and the growing integration of software into hardware, through nanotechnology, may be additional sources of self-expanding power of information processing.

So, a more formal version of this hypothesis is the following: in the first twenty-five years of the information-technology revolution, we have observed a self-generated, expansive capacity of technologies to process information; current limits are likely to be superseded by new waves of innovation in the making; and (this is critical) when and if limits of processing power on the basis of these technologies are reached, a new technological paradigm will emerge—under forms and with technologies that we cannot imagine today, except in the science-fiction scenarios of the usual futurology suspects.

Microelectronics-based technologies are also characterized by their abilities to recombine information in any possible way. This is what I call the hypertext (following the tradition from Nelson to Berners-Lee) and people call the World Wide Web. The real value of the Internet is its ability to link up everything from everywhere and to recombine it. This will be even more explicit when the original design of Berners-Lee's World Wide Web is restored in its two functions, as a browser and as an editor, instead of its current limited uses as a browser/information provider connected to an e-mail system. While Nelson's Xanadu was clearly a visionary utopia, the real potential of the Internet, as Nelson saw it, is in the recombining of all existing information and communication on the basis of specific purposes decided in real time by each user/producer of the hypertext. Recombination is the source of innovation, particularly if the products of this recombination become themselves supports for further interaction, in a spiral of increasingly meaningful information. While the generation of new knowledge will always require the application of theory to recombined information, the ability to experiment with this recombining from a multiplicity of sources considerably extends the realm of knowledge, as well as the connections that can be made between different fields—precisely the source of knowledge innovation in Kuhn's theory of scientific revolutions.

The third feature of new information technologies is their flexibility in allowing the distribution of processing power in various contexts and applications. The explosion of networking technologies (like the Java and Jini languages in the 1990s), the staggering growth of cellular telephony, and the forthcoming full-fledged development of the mobile Internet (that is, cell phone–based access to the Internet from a wide array of portable appliances) are key developments that point to the growing capacity to have processing power, including the power of networked communication, anywhere—anywhere with the technological infrastructure and the knowledge to use it.

I will elaborate more succinctly on the second component of the information-technology revolution, genetic engineering. While it is often considered a process entirely independent from microelectronics, it is not. First, analytically, these technologies are obviously information technologies, since they are focused on the decoding and eventual reprogramming of DNA, the information code of living matter. Second, there is a much closer relationship between microelectronics and genetic engineering than people seem to realize. Without massive computing power and the simulation capacity provided by advanced software, the Human Genome Project would not have been completed—nor would scientists be able to identify specific functions and the locations of specific genes. On the other hand, biochips and chemically based microchips are no longer science-fiction fantasies. Third, there is theoretical convergence between the two technological fields around the analytical paradigm based on networking, self-organization, and emergent properties, as shown in the revolutionary theoretical work of Fritjof Capra.

Genetic-engineering technologies, the transformative powers of which are just being unleashed in the early twenty-first century, are also characterized by their self-expanding processing capacity, by their recombining ability, and by their distributing power. First, the existence of a map of the human genome and, increasingly, of genetic maps of a number of species and subspecies creates the possibility of connecting knowledge about biological processes in a cumulative way, leading to qualitative transformation of our understanding of processes that had been beyond the realm of observation.

Second, the recombining ability concerning DNA codes is exactly what genetic engineering is about and what sets it apart from any previous biological experimentation. But there is a more subtle innovation. The first generation of genetic engineering largely failed because cells were reprogrammed as isolated entities, without understanding that context is everything, in biology as in information processing in general. Cells exist only in their relationships to others. So interacting networks of cells, communicating by their codes rather than by isolated instructions, are the objects of scientific recombination strategies. This kind of recombination is far too complex to be identified in linear terms. It requires simulation techniques with massive computer parallel processing, so that emergent properties are associated with networks of genes, as in some of the models proposed by researchers at the Santa Fe Institute.

Third, the promise of genetic engineering is precisely its ability to reprogram different codes and their protocols of communication, in different areas of different bodies (or systems) of different species. Transgenic research and self-regenerative processes in living organisms are the frontier of genetic engineering. Genetic drugs are intended to induce capabilities of self-programming by living organisms, the ultimate expression of distributed information-processing power.

Incidentally, genetic engineering shows vividly how mistaken we would be to assign positive value to extraordinary technological revolutions independently of their social context, social uses, and social outcomes. I cannot imagine a more fundamental technological revolution than the capacity to manipulate the codes of living organisms. Neither can I think of a more dangerous and potentially destructive technology if it becomes decoupled from our collective capacity to control technological development in cultural, ethical, and institutional terms.

On the foundations of informationalism, the network society emerges and expands throughout the planet as the dominant form of social organization in our time. The network society is a social structure made of information networks powered by the information technologies characteristic of the informationalist paradigm. By *social structure* I mean the organizational arrangements of humans in relationships of production, consumption, experience, and power, as expressed in meaningful interaction framed by culture. A network is a set of interconnected nodes. A node is the point where the curve cuts itself. Social networks are as old as humankind. But they have taken on a new life under informationalism because new technologies enhance the flexibility inherent in networks while solving the coordination and steering problems that impeded networks, throughout history, in their competition with hierarchical organizations. Networks distribute performance and share decision making along the nodes of the network in an interactive pattern. By definition, a network has no center, just nodes. While nodes may be of different sizes, thus of varying relevance, they are all necessary to the network. When nodes become redundant, networks tend to reconfigure themselves, deleting nodes and adding new, productive ones. Nodes increase their importance for the network by absorbing more information and processing it more efficiently. The relative importance of a node does not stem from its specific features but from its ability to contribute valuable information to the network. In this sense, the main nodes are not centers but switches and protocols of communication, following a networking logic rather than a command logic in their performance. Networks work on a binary logic: inclusion/exclusion. As social forms, they are value-free. They can equally kiss or kill: nothing personal. It all depends on the goals of a given network and on its most elegant, economical, and self-reproductive form to perform these goals. In this sense, the network is an automaton. In a social structure, social actors and institutions program the networks. But once programmed, information networks, powered by information technology, impose their structural logic on their human components. That is, until their program is changed—usually at a high social and economic cost.

To apply this formal analysis to the actual workings of society, I will briefly characterize the fundamental structures of this network society.

First of all, the new economy is built on networks. Global financial markets, at the source of investment and valuation, are built on electronic networks processing signals: some of these signals are based on economic calculations, but often they are generated by information turbulences from different sources. The outcomes of these signals, and of their processing in the electronic networks of financial markets, are the actual values assigned to every asset in every economy. The global economy is built around collaborative networks of production and management, as multinational corporations and their ancillary networks account for more than 30 percent of GGP (gross global product) and about 70 percent of international trade. Firms themselves work in and by networks. Large firms are decentralized in internal networks. Small and medium firms form networks of cooperation, thus maintaining their flexibility while pulling resources together. Large firms work on the basis of strategic alliances that vary in products, processes, markets, or periods of time, in a variable geometry of corporate networks. And these corporate networks link up with small and medium business networks, in a world of networks inside networks. Furthermore, what I call the network enterprise often links up customers and suppliers through a proprietary network, as in the business models spearheaded by Cisco Systems or Dell Computer in the electronics industry. The actual operational unit in our economies is the business project, operated by ad hoc business networks. All this complexity can be managed only by the tools of informationalism.

Productivity and competitiveness are vastly enhanced by this networked form of production, distribution, and management. Because networks of the new economy expand throughout the world, phasing out by competition less efficient forms of organization, the new, networked economy becomes the dominant economy everywhere.

Economic units, territories, and people that do not perform well in this economy or that do not present a potential interest for these dominant networks are discarded. On the other hand, any source of potential value, from anywhere and from anything, is connected and programmed into the productive networks of the new economy.

Under such conditions, work is individualized. Management-labor relationships are defined in individual arrangements, and work is valued depending on the capacity of workers or managers to reprogram themselves to perform new tasks and new goals, as the system is driven by technological innovation and entrepreneurial versatility. Not everything is bad in this new working arrangement. It is a world of winners and losers, but, more often than not, of uncertain winners and losers who have no return to the network. It is also a world of creativity as well as of destruction—a world characterized, simultaneously, by creative destruction and destructive creation.

Cultural expression becomes patterned around the kaleidoscope of a global, electronic hypertext. Around the Internet and multimedia, manifestations of human communication and creation are hyperlinked. The flexibility of this media system facilitates the absorption of the most diverse expressions and the customization of the delivery of messages. While individual experiences may exist outside the hypertext, collective experiences and shared messages—that is, culture as a social medium—are by and large captured in this hypertext. It constitutes the source of real virtuality as the semantic framework of our lives. Virtual, because it is based on electronic circuits and ephemeral audiovisual messages. Real, because this is our reality, since the global hypertext provides most of the sounds, images, words, shapes, and connotations that we use in the construction of our meanings in all domains of experience.

Politics is itself increasingly enclosed in the media world, either by adapting to its codes and rules or by attempting to change the rules of the game by creating and imposing new cultural codes. In both cases, politics becomes an application of the hypertext, since the text simply reconfigures itself to the new codes.

Yes, there is life beyond the network society: in the fundamentalist, cultural communes that reject dominant values and build autonomously the sources of their own meaning; sometimes around self-constructed, alternative utopias; more often, around the transcendent truths of God, Nation, Family, Ethnicity, and Territoriality. Thus, the planet is not subsumed entirely by the network society, as the industrial society never extended to the totality of humankind. Yet the networking logic of instrumentality has already linked up dominant segments of societies in most areas of the world around the structural logic embodied in the new, global, networked economy; in the flexible forms of individualized work; and in the culture of real virtuality, inscripted in the electronic hypertext.

The networking logic, rooted in informationalism, has

also transformed our practice of space and time. The space of flows, characteristic of the network society, links up distant locales around shared functions and meanings on the basis of electronic circuits and fast transportation corridors, while isolating and subduing the logic of experience embodied in the space of places. A new form of time, which I call timeless time, emerges out of systemic trends to compress chronological time to its smallest possible expression (as in split-second financial transactions), as well as to blur time sequences, as can be observed in the twisting of professional career patterns away from the predictable progression of the organization man, now replaced by the flexible woman.

Taken into this whirlwind and bypassed by global networks of capital, technology, and information, nation-states do not sink as the prophets of globalization predicted. They adapt in structure and performance, becoming networks themselves. On the one hand, they build supranational and international institutions of shared governance, some of them highly integrated, such as the European Union; others much looser, such as NATO or NAFTA; still others asymmetrical in their obligations, such as the International Monetary Fund, imposing the logic of global markets on developing economies. Yet in all cases, political sovereignty becomes shared among various governments and organizations. On the other hand, in most of the world a process of political decentralization is taking place, shifting resources from national governments to regional and local governments, and even to nongovernmental organizations, in a concerted effort to rebuild legitimacy and increase flexibility in the conduct of public affairs. These simultaneous trends toward supranationality and toward locality induce a new form of state, the network state, which appears to be the most resilient institutional form to weather the storms of the network society.

Where did this network society come from? What was its historical genesis? It emerged from the accidental coincidence of three independent phenomena in the last quarter of the twentieth century.

The first was the information-technology revolution, the key components of which came together as a new technological paradigm in the nineteen-seventies (remember Arpanet, 1969; USENET News, 1979; the invention of the integrated circuit, 1971; the personal computer, 1974–1976; the software revolution: UNIX codes designed in the late sixties, released in 1974; TCP/IP protocols designed in 1973–1978; recombinant DNA, 1973).

The second trend was the process of socioeconomic restructuring of the two competing systems, capitalism and statism, which faced major crises resulting from their internal contradictions in 1973–1975 (capitalism) and 1975–1980 (statism). They both addressed their crises with new government policies and new corporate strategies. The capitalist perestroika worked. The restructuring of statism failed because of the inherent limits of statism to internalize and use the information-technology revolution, as I have argued in my study with Emma Kiselyova of the collapse of the Soviet Union. Capitalism was able to overcome its structural trend toward destructive rampant inflation through informational productivity, deregulation, liberalization, privatization, globalization, and networking, providing the economic foundations of the network society.

The third trend at the origins of this new society was cultural and political and refers to the values projected by the social movements of the late nineteen-sixties and early nineteen-seventies in Europe and in America, with some sui generis manifestations in Japan and China. These movements were fundamentally libertarian, although the feminist movement and the environmental movement extended the notion of freedom to a fundamental challenge to the institutions and ideologies of patriarchalism and productivism. These movements were cultural because they did not focus on the seizing of state power (unlike most of their predecessors in the century) or on redistributing of wealth. Instead, they acted on the categories of experience and rejected established institutions, calling for new meanings of life and, consequently, for the redrawing of the social contracts between the individual and the state and between the individual and the corporate world.

These three phenomena emerged independently from one another. Their historical coincidence was serendipitous, and so was their specific combination in given societies. This is why the speed and shape of the process of transition to the network society is different in the United States, Western Europe, and the rest of the world. The more entrenched the institutions and rules of the industrial society, or of preindustrial societies, the slower and more difficult the process of transformation is. There is no value judgment implied in this differential path toward the network society: the network society is not the promised land of the Information Age. It is, simply, a new, specific social structure, whose effects for the well-being of humankind are undetermined. It all depends on context and process.

One of the key components of this historical accident that produced our twenty-first-century world was the new technological paradigm, informationalism. What was its genesis? War, hot and cold, was an essential ingredient of technological innovation, as has been the case throughout history. World War II was the matrix of most of the discoveries that led to the information-technology revolution. And the cold war was the crucible for their development. Yes, the Internet's ancestor, Arpanet, was not truly a military technology, even if its key technologies (packet switching and distributed networking power) were developed by Paul Baran at Rand Corporation as part of a proposal to the Department of Defense to build a communications system able to survive nuclear war. But the proposal was never approved, and the DOD-based scientists designing Arpanet knew of Baran's work only after they were already building the computer network. However, without the support in resources and freedom of innovation provided by the Advanced Research Projects Agency at the Pentagon, computer science in the United States would not have developed at the pace it did, Arpanet would not have been built, and computer networking would be very different today. Similarly, while the microelectronics revolution has been largely independent of military applications for the last twenty years, in the critical, formative period of the nineteen-fifties and early nineteen-sixties, Silicon Valley and the other major technological centers were highly dependent on military markets and their generous research funding.

Research universities were also essential seedbeds of the technological revolution. In fact, it can be argued that academic computer scientists captured the resources of the Department of Defense to develop computer science in general and computing networking in particular for the sake of scientific discovery and technological innovation, without much direct military application. Actual military design was done under conditions of extreme security in the national laboratories, and there has been very little innovation spun from these laboratories, in spite of their extraordinary scientific potential. They were the mirror of the Soviet system, and so was their fate; they became monumental tombs of ingenuity.

Universities and research centers of major hospitals

and public-health centers were the crucial sources of the biology revolution. Francis Crick and James Watson worked out of Cambridge University in 1953, and the key research leading to the recombinant DNA took place between 1973 and 1975 around Stanford University and the University of California at San Francisco.

Business did play a role, but not so established corporations. AT&T exchanged its proprietary rights to microelectronics for a telecommunications monopoly in the nineteen-fifties and later passed up the opportunity to operate Arpanet in the nineteen-seventies. IBM did not foresee the PC revolution and jumped onto the bandwagon only later, under such confused conditions that it licensed its operating system to Microsoft and left the door open for the PC clones that would end up pushing it to survive mainly as a services company. And as soon as Microsoft became a quasi-monopoly it made similar blunders. It failed to see the Internet's potential until 1995, when it introduced its Internet Explorer, a browser that was not originally created by Microsoft but based on the reworking of a browser designed by Spyglass, a company that licensed Mosaic software from the National Center for Supercomputer Applications. Rank Xerox designed many of the key technologies of the PC age at its PARC research unit in California. But it only half understood the wonders its researchers were doing, so much so that their work was largely commercialized by other companies, particularly by Apple Computer. So the business component at the source of informationalism was, by and large, a new breed of business, start-ups that quickly became giant corporations (Cisco Systems, Dell Computer, Oracle, Sun Microsystems, Apple, etc.) or corporations that reinvented themselves (such as Nokia, which shifted from consumer electronics to cellular telephony and then to the mobile Internet). To be able to transit from their entrepreneurial origins to being innovation-driven, large-scale organizations, these new businesses built on another fundamental component of informationalism: the cultural source of technological innovation represented by the hacker culture.

There are no technological revolutions without cultural transformation. Revolutionary technologies have to be thought of. This is not an incremental process; it is a vision, an act of belief, a gesture of rebellion. To be sure, financing, manufacturing, and marketing will ultimately decide which technologies survive in the marketplace, but not necessarily which technologies develop, because the marketplace, as important as it is, is not the only place in the planet. Informationalism was partly invented and decisively shaped by a new culture that was essential in the development of computer networking, in the distribution of processing capacity, and in the augmentation of innovation potential by cooperation and sharing. The theoretical understanding of this culture and of its role as the source of innovation and creativity in informationalism is the cornerstone in our understanding of the genesis of the network society. In my own analysis, as well as in the contributions of other scholars, this essential dimension of informationalism has been touched upon but not really studied. This is why Pekka Himanen's theory of the hacker culture as the spirit of informationalism is a fundamental breakthrough in the discovery of the world unfolding in this uncertain dawn of the third millennium.

APPENDIX

A Brief History of Computer Hackerism

Now it came to pass that Microsoft had waxed great and mighty among the Microchip Corporations; mightier than any of the Mainframe Corporations before it, it had waxed. And Gates' heart was hardened, and he swore unto his Customers and their Engineers the words of this curse:

"Children of von Neumann, hear me. IBM and the Mainframe Corporations bound thy forefathers with grave and perilous Licenses, such that ye cried unto the spirits of Turing and von Neumann for deliverance. Now I say unto ye: I am greater than any Corporation before me. Will I loosen your Licenses? Nay, I will bind thee with Licenses twice as grave and ten times more perilous than my forefathers. . . . I will capture and enslave thee as no generation has been enslaved before. And wherefore will ye crye then unto

> the spirits of Turing, and von Neumann, and Moore? They cannot hear ye. I am become a greater Power than they. Ye shall cry only unto me, and shall live by my mercy and my wrath. I am the Gates of Hell; I hold the portal to MSNBC and the keys to the Blue Screen of Death. Be ye afraid; be ye greatly afraid; serve only me, and live."[1]

So opens *The Gospel According to Tux,* a hacker "Bible" published on the Web. Tux is the name of the penguin mascot of the Linux computer operating system, created in 1991 by Finnish hacker Linus Torvalds at the age of twenty-two. In the past few years, Linux has attracted a great deal of attention as one of the most serious challengers to Microsoft's primacy.

Anyone may download Linux for free, but this is not the primary difference between Linux and Windows. What distinguishes Linux from the dominant commercial software model epitomized by Microsoft's products is first and foremost its openness: in the same way scientific researchers allow all others in their fields to examine and use their findings, to be tested and developed further, hackers who take part in the Linux project permit all others to use, test, and develop their programs. In research, this is known as the scientific ethic. In the field of computer programming, it is called the open-source model ("source code" being a program's DNA, its form in the language used by programmers to develop it; without the source code, a person can use a program but is not able to develop it in new directions).

This kinship with the academic research model is not accidental: openness may be seen as a legacy that hackers have received from the university. *The Gospel According to Tux* elevates to heroic status the researchers who openly shared their findings while creating the theoretical foundation for the computer, chief among them Alan Turing and John von Neumann.

Optimistically, *The Gospel According to Tux* goes on to relate how Torvalds revives this spirit in the world of computers:

> Now in those days there was in the land of Helsinki a young scholar named Linus the Torvald. Linus was a devout man, a disciple of RMS [Richard Stallman, another famous hacker] and mighty in the spirit of Turing, von Neumann and Moore. One day as he was meditating on the Architecture, Linus fell into a trance and was granted a vision. And in the vision he saw a great Penguin, serene and well-favoured, sitting upon an ice floe eating fish. And at the sight of the Penguin Linus was deeply afraid, and he cried unto the spirits of Turing, von Neumann and Moore for an interpretation of the dream.
>
> And in the dream the spirits of Turing, von Neumann and Moore answered and spoke unto him, saying, "Fear not, Linus, most beloved hacker. You are exceedingly cool and froody. The great Penguin which

> you see is an Operating System which you shall create and deploy unto the earth. The ice floe is the earth and all the systems thereof, upon which the Penguin shall rest and rejoice at the completion of its task. And the fish on which the Penguin feeds are the crufty Licensed code bases which swim beneath all the earth's systems.
>
> The Penguin shall hunt and devour all that is crufty, gnarly and bogacious; all code which wriggles like spaghetti, or is infested with blighting creatures, or is bound by grave and perilous Licenses shall it capture. And in capturing shall it replicate, and in replicating shall it document, and in documentation shall it bring freedom, serenity and most cool froodiness to the earth and all who code therein.

Linux did not invent the open-source model, nor did it appear out of nowhere. Linux is a Unix-like operating system built on the foundation of two earlier hacker projects. Most important to Linux was the GNU operating-system project begun by Richard Stallman in 1983.[2] Stallman, who started out in MIT's AI Lab, continues to work in the tradition of that first nexus of hackerism.

The other matrix for Linux is BSD Unix, created by Bill Joy in 1977. BSD stands for Berkeley Software Distribution, in homage to its origins at another traditional hacker center, the University of California at Berkeley, where Joy started developing his operating system when he was a graduate student in his twenties.[3]

Another important chapter in the history of computer hackerism came with the birth of the Internet. Its true beginnings date back to 1969 (which was also when hackers Ken Thompson and Dennis Ritchie wrote the very first version of Unix).[4] The U.S. Department of Defense's research unit ARPA (Advanced Research Projects Agency) played an important role in setting up the Internet's predecessor, the Arpanet. However, the extent and significance of this governmental input is usually exaggerated.[5] In *Inventing the Internet*, the most thorough history of the Internet to date, Janet Abbate demonstrates how the appointment of former university researchers to managerial positions caused the Internet to develop according to self-organizing principles common to scientific practice. As a result, the most significant portion of that development was soon directed by the Network Working Group, a cluster of hackers culled from a talented group of university students.

The Network Working Group operated on the open-source model: anyone was allowed to contribute ideas, which were then developed collectively. The source codes of all solutions were published from the very beginning, so that others could use, test, and develop them. This model is still followed. The composition and name of this spearheading hacker group has changed many times along the way. Currently it is known as the Internet Engineering Task Force, and it operates under the Internet Society founded by Vinton Cerf, a charter member of the group from his days as a graduate student in computer science at UCLA. Cerf has played an important role in almost all the technological advancements in the evolution of the Net. One aspect has always remained the same, however: the Internet does not have any central directorate that guides its development; rather, its technology is still developed by an open community of hackers.[6] This community discusses ideas, which become "standards" only if the larger Internet community thinks they are good and starts to use them. Sometimes these hacker ideas have taken the Net in totally unanticipated directions, such as when Ray Tomlinson introduced e-mail in 1972. (He chose the @ symbol we still use in e-mail addresses.) Reflecting on this development, Abbate notes that "there seems to have been no corporate participation in the design of the Internet. Like its predecessor [the Arpanet], the Internet was designed, informally and with little fanfare, by a self-selected group of experts."[7]

Nor was the World Wide Web, the global hypertext built on the basis of the Internet, a corporate or governmental construction. Its prime mover was an Oxford-educated Englishman, Tim Berners-Lee, who started planning the design of the Web in 1990 while working at the Swiss particle-physics research center CERN. Behind Berners-Lee's unassuming exterior, he is a strong idealist who remains outspoken about his vision of how the Web can make this a better world: "The Web is more a social creation than a technical one. I designed it for a social effect—to help people work together—and not as a technical toy. The ultimate goal of the Web is to support and improve our weblike existence in the world."[8]

Gradually, other hackers joined him in this effort, as he describes in his book *Weaving the Web* (1999): "Interested people on the Internet provided the feedback, stimulation, ideas, source-code contributions, and moral support that would have been hard to find locally. The people of the Internet built the Web, in true grassroots fashion."[9] As the group expanded, Berners-Lee organized a community similar to Cerf's Internet Society, the World Wide Web Consortium, in an effort to forestall a commercial takeover of the Web. Personally, Berners-Lee has resolutely refused all commercial offers, which one of his friends has characterized as typical of his general outlook: "As technologists and entrepreneurs were launching or merging companies to exploit the Web, they seemed fixated on one question: 'How can I make the Web mine?' Meanwhile, Tim was asking, 'How can I make the Web yours?' "[10]

The most important individual behind the Web's final breakthrough was Marc Andreessen, who studied at the University of Illinois at Champaign-Urbana. In 1993, at the university's National Center for Supercomputing Applications, the twenty-year-old Andreessen and a few other hackers created a user-friendly graphical browser for the PC. This program, distributed with open source code, soon led to the even better known and more rapidly disseminated Netscape Navigator browser.[11]

Although at the moment the Internet and the Web (together "the Net") dominate our collective imagination, their mass breakthrough would not have been possible, of course, without the creation of that other remarkable invention of our time, the personal computer. Its ideational history goes back to the first MIT hackers who pioneered interactive computing. In their time, the computer field was still dominated by IBM's model of batch-processed mainframe computers, in which programmers did not have direct access to the computer but had to receive permission to pass their programs on to a special operator. It could take days to receive the results. In contrast to this method, the MIT hackers favored interactive computing on minicomputers, in which the programmer could write his program directly into the computer, see the results, and immediately make desirable corrections. In terms of social organization, the difference is great: in an interaction that eliminates the "operator," individuals can employ technology in a more liberating manner. This elimination of the operators, the high priesthood of the computer world, is experientially comparable to the elimination of telephone operators in the history of the telephone. It meant a freeing up of direct exchange between individuals.[12]

The MIT hackers also programmed the first ever computer game, in which a user could for the first time experience the possibilities of the graphical user interface. In Steve Russell's 1962 *Spacewar,* two vessels armed with torpedoes, guided by controls designed by the club, joined battle in outer space. Peter Samson added a planetary background to the game, called "Expensive Planetarium" because its purpose was to show the stars in exactly the same positions they could have been seen by looking out the window—but much more expensively, as user time on the computer was very valuable back then. Anyone was allowed to copy the game, and its source code was available.[13]

The final breakthrough of the personal computer was made possible by these mental preparations. The decisive further step was taken by Steve Wozniak, who was a member of the Homebrew Computer Club, a group of hackers who started meeting regularly in the Bay Area in the mid-seventies. In 1976, using the information shared freely within the club, he built, at the age of twenty-five, the first personal computer for the use of people without engineering degrees, the Apple I. To appreciate the importance of this accomplishment, we must remember that the computers preceding it had often been machines the size of refrigerators that had to be kept in special climate-controlled rooms. The CEOs of the world's largest computer firms did not believe in a future for personal computers, expressing opinions such as "I think there is a world market for maybe five computers" (Thomas Watson, President of IBM, 1943) and "There is no reason anyone would want a computer in their home" (Ken Olsen, cofounder and chairman of Digital Equipment Corporation, 1977). These predictions might even have come true if Woz had not succeeded in "humanizing" the computer.

Woz's achievement in making the computer available to

everyone reflected the Bay Area's overall countercultural spirit and its concern with empowering people in various ways. Just before Woz made his first computer, Ted Nelson, a visionary whose charisma can make him seem like a frenzied shaman, heralded the coming of the personal computer in a self-published book called *Computer Lib* (1974). Nelson is best known for expressing a vision of a worldwide hypertext long before the advent of the Web, and he is in fact the inventor of the term *hypertext.* In his book, his rallying cry was "COMPUTER POWER TO THE PEOPLE! DOWN WITH CYBERCRUD." (*Cybercrud* is a term Nelson coined to refer to ways of "putting things over on people using computers.")[14]

Later on, Woz himself stressed that the atmosphere of the Homebrew Computer Club, which Nelson visited, energized him in his work on the Apple I: "I came from a group that was what you might call beatniks or hippies—a lot of technicians who talked radical about a revolution in information and how we were going to totally change the world and put computers in homes."[15] In accord with the hacker ethic, Woz openly distributed blueprints of his computer to others and published bits of his program. His hacker-created computer inspired the larger personal-computer revolution, the consequences of which are everywhere around us.[16]

Notes

Preface

1. *The Jargon File,* s.v. *hacker.* This file is maintained by Eric Raymond at www.tuxedo.org/~esr/jargon. It has also been published as *The New Hacker's Dictionary* (3d ed., 1996).
2. *The Jargon File,* s.v. *hacker ethic.*
3. In *Hackers: Heroes of the Computer Revolution* (1984), Levy describes the spirit of the MIT hackers by saying that they believed that "all information should be free" and that "access to computers . . . should be unlimited and total" (p. 40).
4. *The Jargon File* gives this definition for *cracker:* "One who breaks security on a system. Coined ca. 1985 by hackers in defense against journalistic misuse of hacker." It is worth noting that in his 1984 book on hackers, Levy did not yet see any need to describe the difference between hackers and crackers. This is related to the fact that the history of computer viruses, or self-propagating computer programs, really began in the second half of the eighties. The concept "computer virus" itself gained currency from Fred Cohen's dissertation on the subject in 1984, and the first real-world PC viruses were spread by means of diskettes in 1986 (cf. Solomon, "A Brief History of PC Viruses" [1990] and Wells, "Virus Timeline" [1996]). The first notorious example, of break-ins into information systems also took place in the latter half of the eighties. One of the most fa-

mous cracker groups, Legion of Doom, was founded in 1984, and the cracker manifesto of a later member of the group, the Mentor, was published in 1986 ("The Conscience of a Hacker," in which it is noteworthy that "crackers" started to call themselves "hackers"; for the history of the group, cf. "The History of the Legion of Doom" [1990]).

5. *The Jargon File*, s.v. *hacker.*
6. *Die protestantische Ethik und der Geist des Kapitalismus* in *Archiv für Sozialwissenschaft und Sozialpolitik*, vols. 20–21 (1904–1905), which was reprinted in revised form in the series *Gesammelte Aufsätze zur Religionssoziologie* (1920).

Chapter 1: The Hacker Work Ethic

1. Hafner and Lyon, *Where Wizards Stay Up Late: The Origins of the Internet* (1998), p. 139.
2. Wolfson and Leyba, "Humble Hero."
3. Flannery with Flannery, *In Code: A Mathematical Journey* (2000), p. 182.
4. A message to comp.os.minix on December 19, 1991.
5. Berners-Lee, *Weaving the Web*, pp. 9–13.
6. Connick, ". . . And Then There Was Apple" (1986), p. 24.
7. Flannery, *In Code*, p. 182.
8. Raymond, "The Art of Unix Programming" (2000), chap. 1.
9. Letter 7.341c–d. This academic passion is a persistent theme in all of Plato's Socratic writings. In the *Symposium,* Plato has Alcibiades speak of the "Bacchic frenzy of philosophy" transmitted to him by Socrates (218b). In the *Phaedrus,* this notion is extended by the statement that common people regard philosophers as madmen but that theirs is a divine madness (or higher passion). Plato also emphasizes the literal meaning of the word *philosophy* ("a passion or love for wisdom") in the dialogues concerned with the role of philosophy: *Republic, Symposium, Phaedrus, Theaetetus,* and *Gorgias,* as well as in the *Apology.*
10. Levy, *Hackers*, p. 434.
11. Raymond, "How to Become a Hacker" (1999), p. 232.
12. *Die protestantische Ethik und der Geist des Kapitalismus,* in *Archiv für Sozialwissenschaft und Sozialpolitik,* vols. 20–21 (1904–1905), which was reprinted in revised form in the series *Gesammelte Aufsätze zur Religionssoziologie* (1920).

13. Weber, *The Protestant Ethic and the Spirit of Capitalism,* pp. 54, 61–62.
14. Baxter, *Christian Directory,* cited in Weber, *Protestant Ethic,* p. 157, n. 9, and p. 158, n. 15.
15. Ibid., p. 177, n. 101.
16. Ibid., p. 158, n. 18.
17. *The Rule of St. Benedict,* 48.
18. Cassian, *The Twelve Books on the Institutes of the Coenobia,* 4.26.
19. The famous hermit Anthony, who is considered the founder of Christian monasticism in the fourth century, set an example for the later monastic movement by working. Athanasius describes him in his *Life of Anthony:* "He worked, however, with his hands, having heard, 'he who is idle let him not eat' [2 Thess. 3:10], and part he spent on bread and part he gave to the needy" (3). See also *Apophthegmata Patrum:*

> When the holy Abba Anthony lived in the desert he was beset by *accidie* [the restlessness of the soul], and attacked by many sinful thoughts. He said to God, "Lord, I want to be saved, but these thoughts do not leave me alone; what shall I do in my affliction? How can I be saved?" A short while afterwards, when he got up to go out, Anthony saw a man like himself sitting at his work, getting up from his work to pray, then sitting down and plaiting a rope, then getting up again to pray. It was an angel of the Lord sent to correct and reassure him. He heard the angel saying to him, "Do this and you will be saved." At these words, Anthony was filled with joy and courage. He did this, and he was saved. (Anthony I, trans. in Ward, ed., *The Sayings of the Desert Fathers* [1975])

In addition to the monastic rules of Cassian and Benedict, Basil's rule was important. He talks about how working makes one chaste:

> Our Lord Jesus Christ says: "He is worthy" not everyone without exception or anyone at all, but "the workman, of his meat" [Mt. 10:10] and the Apostle bids us labor and work with our own hands the things which are good, that we may have something to give to him that suffereth need. It is, therefore, immediately obvious that we must toil with diligence and not think that our goal of piety offers an escape from work

> or a pretext for idleness, but occasion for struggle, for ever greater endeavor, and for patience in tribulation, so that we may be able to say: "In labor and painfulness, in much watchings, in hunger and thirst" [2 Cor. 11:27]. (*The Long Rules,* 37)

The only ancient philosophy that praised work was Stoicism, the influence of which on monastic thinking is well known. For example, Epictetus taught: "Ought we not, as we dig and plough and eat, to sing the hymn of praise to God?" and "What then? Do I say that man is an animal made for inactivity? Far be it from me!" (*Discourses,* 1.16 and 1.10). Naturally, the monks and Stoics did not go as far as the Protestant ethic in their appreciation of work, as Birgit van den Hoven shows in her research, *Work in Ancient and Medieval Thought* (1996).

20. Benedict writes: "But if anyone of them [the working monks] should grow proud by reason of his art, in that he seemeth to confer a benefit on the monastery, let him be removed from that work and not return to it, unless after he hath humbled himself, the Abbot again ordereth him to do so" (*The Rule of St. Benedict,* 57).
21. Weber, *Protestant Ethic,* pp. 181–83. Weber's study has two dimensions. On one hand, it is the historical proposition that the Protestant ethic had an important influence on the formation of the spirit of capitalism. On the other, it is the suprahistorical thematization of a certain social ethic. Since the first one of these dimensions is, to some extent, empirically questionable—for example, the same capitalist spirit evolved also in contemporary Catholic Venice (a brief summary of the other main counterarguments by Anthony Giddens in his introduction to Weber's English translation)—and is no longer an essential factor in the consideration of our own time, I will focus on the second one, using the terms *the spirit of capitalism* and *Protestant ethic* thematically, not historically. Since their two main points are the same, they may be used interchangeably in a thematic discussion. (For more, see Weber's characterization of the relation between the Protestant ethic and the spirit of capitalism, pp. 54–55, 72, 91–92.)
22. Castells, *Information Age* (2000), 1:468. Martin Carnoy concludes similarly in his *Sustaining the New Economy: Work, Family, and Community in the Information Age* (2000): "The absence of a relation between IT industry and employment growth or unemployment suggests that the level of unemployment is a result of factors other than the rate of IT diffusion" (p. 38).

23. Augustine, *Concerning the City of God Against the Pagans,* 22.30. According to Augustine, "We ourselves shall become that seventh day, when we have been replenished and restored by his blessing and sanctification" (ibid.). Gregory the Great wrote in the sixth century:

> The actual Passion of our Lord and His actual resurrection prefigure something about His mystical Body in the days of its passion. On Friday He suffered, on Saturday He rested in the tomb, on Sunday He arose from death. The present life is to us as Friday, because it is led amid sorrows and beset with difficulties. But on Saturday, as it were, we rest in the tomb, because we find rest for the soul after it has been freed from the body. On Sunday, however, the third day from the Passion, or as we have said, the eighth day from the beginning of time, we will rise bodily from the dead and we will rejoice in the glory of the soul together with the body. (*Homilies on the Book of the Prophet Ezechiel,* 2.4.2)

24. *On Genesis Against the Manichees,* 2.11.
25. When Tundale in his vision took his tour of the beyond guided by an angel, he saw in a place called Vulcan how malefactors were being tortured with hammers and other tools. His ears were filled with the frightful noise of hammers on anvils, etc., and the traditional energy source of labor, fire, was scorching the sinners:

> They seized the soul who followed, and holding onto him they threw him into the burning forge, its flames fanned with inflated bellows. Just as iron is usually weighed, these souls were weighed, until the multitude that were burned there was reduced to nothing. When they were so liquefied that they appeared to be nothing but water, they were thrown with iron pitchforks. Then placed on a forging stone they were struck with hammers until twenty or thirty or a hundred souls were reduced into one mass. ("Tundale's Vision," in Gardiner, ed., *Visions of Heaven and Hell Before Dante* [1989], pp. 172–73)

Eileen Gardiner comments aptly on the vision literature's image of Hell:

> Awful smells and horrendous noise are associated with hell, along with other assaults on the tactile and visual senses. Hell

> is clearly imagined and described over and over. Often the details are the same—fire, bridges, burning lakes, horrid little creatures pulling out sinners' entrails. They are physical, colourful, vivid images. They are often related to the masculine images of work provided by the nascent industrial economy. Forges, furnaces, hammers, smoke, and burning metals combine to present a picture that would certainly be hellish to a rural, aristocratic, or agrarian audience. (*Medieval Visions of Heaven and Hell: A Sourcebook* [1993], p. xxviii)

26. "St. Brendan's Voyage," in Gardiner, *Visions of Heaven and Hell,* pp. 115–16.
27. Dante, *The Divine Comedy*, *Inferno,* 7.25–35.
28. Homer: "Yes, and I saw Sisyphus in bitter torment, seeking to raise a monstrous stone with his two hands. In fact he would get a purchase with hands and feet and keep pushing the stone toward the crest of a hill, but as often as he was about to heave it over the top, the weight would turn it back, and then down again to the plain would come rolling the shameless stone. But he would strain again and thrust it back, and the sweat flowed down from his limbs, and dust rose up from his head" (*Odyssey,* 11.593–600). The horrors of the Sisyphean labors are also mentioned by Plato in *Gorgias,* 525e (cf. also *Apology,* 41c, and *Axiochus,* 371e).
29. Lavater, *Aussichten in die Ewigkeit* (1773), 3:93.
30. Ulyat, *The First Years of the Life of the Redeemed After Death* (1901), p. 191.
31. Defoe, *Robinson Crusoe,* pp. 211–12. This is how Crusoe describes the need for counting time:

> After I had been there about ten or twelve days, it came into my thoughts that I should lose my reckoning of time for want of books and pen and ink, and should even forget the Sabbath days from the working days; but to prevent this I cut it with my knife upon a large post, in capital letters, and making it into a great cross I set it up on the shore where I first landed, viz. 'I came on shore here on the 30th of Sept. 1659.' Upon the sides of this square post I cut every day a notch with my knife, and every seventh notch was as long again as the rest, and every first day of the month as long again as that long one,

> and thus I kept my kalander, or weekly, monthly, and yearly reckoning of time. (p. 81)

But this habit of taking Sundays off was forgotten very soon (p. 89).

32. Tournier, *Friday* (1967/1997), p. 151.
33. Crusoe is an excellent example of our changed attitudes to work, because the idea of living on an island is an apt illustration of our values.

 Crusoe's life on his island is very different from that depicted in the ancient myth of the Islands of the Blessed, where, according to Hesiod, people live in a state similar to that of the Golden Age, when people "lived like gods without sorrow of heart, remote and free from toil and grief: miserable age rested not on them; but with legs and arms never failing they made merry with feasting beyond the reach of all evils" (*Work and Days,* 114–17).

 Images of life on an island have also influenced the history of utopias, and the difference between the ancient and modern conceptions is very clear. Socrates' (i.e., Plato's) ideal society was modeled on the Island of the Blessed. In the best possible society, only the lowest classes and slaves would work. Socrates explains: "There are other servants, I think, whose mind alone wouldn't qualify them for membership in our society but whose bodies are strong enough for labor. These sell the use of their strength for a price called a wage and hence are themselves called wage-earners. Isn't that so?" (*Republic,* 371d–e; cf. also 347b, 370b–c, 522b, 590c). The citizens, in the word's full sense, are free from work and devote their time to philosophy. This Socratic relation to work is strong in all of Plato's writings. In *Gorgias,* Plato makes Socrates comment to his interlocutor Callicles that as a free man he certainly would not let his daughter marry an engineer, and "[you'd] despise him and his craft, and you'd call him 'engineer' as a term of abuse" (512c; cp. 518e–19a). In *Phaedrus,* Socrates even presents a "top-ten list" of fates in life. Only being a sophist, a tyrant, and an animal are listed lower than being a worker (not surprisingly, the number-one spot goes to gods or godlike humans—i.e., philosophers) (248d–e). The tone is the same in other Platonic writings, too (cf. especially *Symposium,* 203a, and *Alcibiades,* 1:131b).

 The attitudes toward work in modern utopias have been strikingly different. On Thomas More's island of Utopia, idleness is in fact prohibited—an idea that has been shared by most of the other well-known utopias envisioned since the Renaissance.

34. A message to comp.os.minix on January 29, 1992.
35. Raymond, "How to Become a Hacker," p. 233.
36. Ibid., p. 237.

Chapter 2: Time Is Money?

1. The whole passage from *Advice to a Young Tradesman* (1748) reads: "Remember, that *time* is money. He that can earn ten shillings a day by his labour, and goes abroad, or sits idle, one half of that day, though he spends but sixpence during his diversion or idleness, ought not to reckon *that* the only expense; he has really spent, or rather thrown away, five shillings besides" (p. 370).
2. Cf. *Information Age* (2000), vol. 1, chap. 7. *Informational economy* also means an economy whose characteristic products are information technology or information itself. Pine and Gilmore add an important level when they talk about the new *experience economy.* The informational economy is also an economy of symbols, in which the symbolic level of the products becomes more and more important. Pine and Gilmore write about this economy's consumer: "When he buys an experience, he pays to spend time enjoying a series of memorable events that a company stages—as in a theatrical play—to engage him in a personal way" (*The Experience Economy* [1999], p. 2). Even if the consumer is not conscious that he or she wants to consume an experience when drinking a cup of coffee in a café of a certain style, companies more and more consciously design their products as experiences because that sells.
3. Ibid., vol. 1, chap. 2. The empirical data is also provided by Held et al., eds., *Global Transformations: Politics, Economics, and Culture* (1999).
4. Spector, *Amazon.com: Get Big Fast* (2000), p. 41.
5. Moore first presented his law in the "Experts Look Ahead" series in *Electronics* magazine in 1965. According to its initial formulation, the number of components that can be packed into an integrated circuit doubles every year. Later, this figure was corrected to every eighteen months. The law is sometimes expressed in a more easily remembered form: every other year, efficiency doubles and cost is cut in half.
6. Clark with Edwards, *Netscape Time: The Making of the Billion-Dollar Start-Up That Took on Microsoft* (1999), pp. 67–68, 62–63.
7. Cf. also *Information Age* (2000), vol. 1, chap. 3.

8. Ibid., chap. 4.
9. Michael Dell, the founder of Dell Computer, has expressed this principle of networking succinctly in his "rules for Internet revolutionaries": "Turn over to outsiders operations that aren't central to your business." He goes on to say, "Choose what you want to excel at, and find great partners for the rest." Dell with Fredman, *Direct from Dell: Strategies That Revolutionized an Industry* (1999), pp. xii, 173.
10. Hammer has explained his theory in a more popularized form with James Champy in *Reengineering the Corporation* (1993). In it, he discusses the questions that successful organizations pose to themselves: "They weren't asking: 'How can we do what we do better?' or 'How can we do what we do at a lower cost?' Instead, they were asking 'Why do we do what we do at *all*?' " Looking at management through this question, Hammer and Champy concluded: "We found that many tasks that employees performed had nothing at all to do with meeting customer needs—that is, creating a product high in quality, supplying that product at a fair price, and providing excellent service. Many tasks were done simply to satisfy the internal demands of the company's own organization" (p. 4). Instead of this, Hammer and Champy urge companies to organize around the key process.
11. Dell summarizes this principle: "Velocity, or the compression of time and distance backward into the supply chain and forward to the customer, will be the ultimate source of competitive advantage. Use the Internet to lower the cost of developing links between manufacturers and suppliers, and manufacturers and customers. This will make it possible to get products and services faster to market than ever before." Dell, *Direct from Dell*, p. xii.
12. Rybczynski, *Waiting for the Weekend*, p. 18. It is appropriate that the first person known to no longer merely play tennis but systematically work on his backhand was none other than Frederick Taylor. To this end, he even designed a special racquet and won the U.S. men's doubles championship in 1881. Copley, *Frederick W. Taylor: Father of the Scientific Management*, 1:117.
13. Kantrowitz, "Busy Around the Clock" (2000), p. 49.
14. Russell Hochschild, *Time Bind* (1997), p. 209. This has actually realized the larger vision that Taylor expressed in the introduction to his book: "The same principles [of scientific management] can be applied with equal force to all social activities." He mentions "the management of our homes" as the first example (p. iv).

15. Ibid., p. 232.
16. Ibid., p. 50.
17. Weber, *Protestant Ethic*, p. 161.
18. Aronson and Greenbaum, "Take Two Aspirin," typescript. Cited in Fischer, *America Calling* (1992), p. 176.
19. Fischer, *America Calling*, photo 7.
20. Ibid., photo 8.
21. Plato, *Theaetetus*, 172d; cf. 172c–73b, 154e–55a, and 187d–e. See also *Apology*, 23c, and *Phaedrus*, 258e.
22. Le Roy Ladurie, *Montaillou* (1978), pp. 279, 277.
23. Ibid., pp. 277–78.
24. *The Rule of St. Benedict*, 18.
25. Ibid., 16.
26. Ibid., 42.
27. Ibid., 11.
28. Ibid., 43.
29. In fact, the latecomers are appropriately punished punctually at certain times: "Then, at all Hours, when the Work of God is ended, let him cast himself on the ground in the place where he standeth, and thus let him make satisfaction, until the Abbot again biddeth him finally to come from this penance" (ibid., p. 44).
30. Benedict writes: "But in the nocturnal assemblies a late arrival up to the *second* Psalm is allowed, provided that before the Psalm is finished and the brethren bow down in prayer he makes haste to take his place in the congregation and join them; but he will most certainly be subjected to the same blame and penance which we mentioned before if he has delayed ever so little beyond the hour permitted for a late arrival."
31. Franklin, *Autobiography*, p. 90.
32. Thompson also wrote the book *The Making of the English Working Class* (1963) on the theme.
33. Brand, *The Media Lab*, p. 53.
34. Raymond, "How to Become a Hacker," p. 236.

Chapter 3: Money as a Motive

1. Weber, *Protestant Ethic*, p. 53.
2. *The Jargon File*, s.v. *hacker ethic*.
3. *The Rule of St. Benedict*, 6.
4. Tertullian puts this succinctly: "Restless curiosity, the feature of heresy" (*Presciption Against Heretics*, 14).

5. Merton's classic article "Science and Technology in a Democratic Order" (*Journal of Legal and Political Sociology* 1 [1942]) has been reprinted as "The Normative Structure of Science" in his collection *The Sociology of Science: Theoretical and Empirical Investigations* (1973). Cf. pp. 273–75.
6. The significance of *synusia* is discussed in Plato's Letter 7. Research indicates that the common image of Plato's Academy, like that in Raphael's pompous painting *The School of Athens,* does not correspond to historical facts. The Academy does not seem to have been so much a university building or campus in the modern sense but much more a certain philosophy of science that loosely linked people. The Academy was a group of scholars that met in a park outside Athens's city limits, called the Akademeia, after the Athenian hero Akademos. To state, as some writings of antiquity do, that Plato bought this park is as absurd as the claim that someone today could go ahead and purchase New York's Central Park or simply announce that she or he intended to build a private university there. Plato may well have owned a house close to the park. Cf. Baltes, "Plato's School, the Academy" (1993); Cherniss, *The Riddle of the Early Academy* (1945); Dillon, "What Happened to Plato's Garden?" *Hermathena* (1983); Glucker, *Antiochus and the Late Academy* (1978); Dusanic, "Plato's Academy and Timotheus' Policy, 365–359 B.C." (1980); Billot, "Académie" (1989); and Gaiser, *Philodems Academica: die Bericht über Platon und die Alte Akademie in zwei herkulanensischen Papyri* (1988).

   Similarly, Ficino's academy, which revived the Platonic Academy, seems not to have been a physical building but a revival of this philosophy of science. Cf. Hankins, "The Myth of the Platonic Academy of Florence" (1991).
7. Stallman, "The GNU Operating System and the Free Software Movement" (1999), p. 59n. For a description of other forms of open-source licenses, see Perens, "The Open Source Definition" (1999), which is updated at www.opensource.org/osd.html.
8. Gold, *Steve Wozniak: A Wizard Called Woz* (1994), p. 10.
9. Aristotle writes: "There still remains one more question about the citizen: Is he only a true citizen who has a share of office, or is the mechanic to be included? . . . It must be admitted that we cannot consider all those to be citizens who are necessary to the existence of the state. . . . The best form of state will not admit them [the artisans] to citizenship" (*Politics,* 1277b–78a).
10. Raymond, "Homesteading the Noosphere" (1998), p. 100.

11. Brand, *Media Lab*, p. 57.
12. *The Importance of Living*, p. 158. He adds: "The danger is that we get over-civilized and that we come to a point, as indeed we have already done, when the work of getting food is so strenuous that we lose our appetite for food in the process of getting it."
13. Linzmayer, *Apple Confidential* (1999), pp. 37–40.
14. Wolfson and Leyba, "Humble Hero."
15. Southwick, *High Noon: The Inside Story of Scott McNealy and the Rise of SUN Microsystems* (1999), p. 16. For the history of the company's founding, see chap. 1.
16. Ceruzzi, *A History of Modern Computing* (1998), chap. 7. Among Microsoft's first languages were BASIC (1975), FORTRAN (1977), and COBOL-80 (1978). From the later perspective of Microsoft's attacks on Unix-like operating systems (the latest case being the attacks on Linux in the internal memorandums that were leaked to the public: Valloppillil, *Open Source Software* [1998]; Valloppillil and Cohen, *Linux OS Competitive Analysis* [1998]), it is a bit ironic that its first operating system was also a version of the hacker-favored Unix-family, XENIX ("Microsoft Timeline").
17. Gates, *The New York Times Syndicate.*
18. The story of Red Hat is described in Young with Goldman Rohm, *Under the Radar* (1999).
19. Stallman, "The Free Software Song."
20. "What Is Free Software?" (1996). For other serious treatments of the topic, see "The GNU Manifesto" (1985) and "The GNU Operating System and the Free Software Movement" (1999).
21. This is the difference between those who prefer Stallman's *free software* and those who prefer *open source.* One of the reasons for adopting the new term suggested by Chris Peterson in a meeting of a few leading hackers in Palo Alto in February 1998 was to be less ideological. The two most famous proponents of this new term are Bruce Perens and Eric Raymond, who founded opensource.org for spreading the idea. Cf. Opensource.org, "History of the Open Source Initiative." Cf. also Rosenberg, *Open Source: The Unauthorized White Papers* (2000) and Wayner, *Free for All: How Linux and the Free Software Movement Undercut the High-Tech Titans* (2000).
22. Anthony, *The Ideology of Work* (1977), p. 92.

Chapter 4: The Academy and the Monastery

1. Weber, *Protestant Ethic,* p. 64.
2. *The Jargon File,* s.v. *hacker ethic.*
3. For the early history, see Torvalds, "Re: Writing an OS" (1992) and "Birthday" (1992).
4. Cf. Tanenbaum, *Operating Systems: Design and Implementation* (1987).
5. Torvalds, "What Would You Like to See Most in Minix?" (1991).
6. Torvalds, "Birthday" (1992).
7. On October 5, 1991, Torvalds posted a message asking, "Do you pine for the nice days of minix-1.1, when men were men and wrote their own device drivers?" Torvalds, "Free Minix-like Kernel Source for 386-AT" (1991).
8. For a more comprehensive look at the contributors to the Linux project, see Torvalds, "Credits," and Dempsey, Weiss, Jones, and Greenberg, *A Quantitative Profile of a Community of Open Source Linux Developers* (1999).
9. The first discussion took place within the newsgroup comp.os.minix. Linux 0.0.1 became available on the Finnish server nic.funet.fi in the directory /pub/OS/Linux in September 1991. Nowadays, Torvalds uploads the newest version of the kernel to ftp.kernel.org/pub/linux/kernel. There are countless mailing lists, newsgroups, and webpages focused on Linux.
10. Raymond writes:

> The most important feature of Linux, however, was not technical but sociological. Until the Linux development, everyone believed that any software as complex as an operating system had to be developed in a carefully coordinated way by a relatively small, tightly knit group of people. This model was and still is typical of both commercial software and the great freeware cathedrals built by the Free Software Foundation in the 1980s—also of the freeBSD/netBSD/OpenBSD projects that spun off from the Jolitzes' original 386BSD port.
>
> Linux evolved in a completely different way. From nearly the beginning, it was rather casually hacked on by huge numbers of volunteers coordinating only through the Internet. Quality was maintained not by rigid standards or autocracy but by the naively simple strategy of releasing every week and

> getting feedback from hundreds of users within days, creating a sort of rapid Darwinian selection on the mutations introduced by developers. ("The Cathedral and the Bazaar" [1999], pp. 23–24)

11. Merton, "Normative Structure of Science," in *Sociology of Science.* Cf. p. 277.
12. Basically, all of Plato's Socratic dialogues are examples of this critical dialogue; in them, Socrates often makes remarks about the need for critical dialogue. For example, in *Crito,* Socrates says, "Let us examine the question together, my dear friend, and if you make any objection while I am speaking, make it and I will listen to you" (48e). In *Phaedo,* he incites his interlocutor to criticize him by asking, "Do you think there is something lacking in my argument?" and in *Euthydemus* he remarks similarly: "There is nothing I would like better than to be refuted on these points" (295a). In *Theaetetus* and *Clitophon,* Socrates explains why the process of critique is always beneficial: "Either we shall find what we are going out after; or we shall be less inclined to think we know things which we don't know at all—and even that would be a reward we could not fairly be dissatisfied with" (187b–c); and, "Once I know my good and bad points, I will make it my practice to pursue and develop the former while ridding myself of the latter to the extent that I am able" (407a). For this reason, in academic discussion one should present a critique frankly and not try to please anyone (cf. *Euthyphro,* 14e; *Protagoras,* 319b, 336e; *Republic,* 336e).
13. In fact, William Whewell, who coined the word *scientist* in the nineteenth century, meant by that word a person participating in such a self-corrective process.
14. Kuhn said that paradigms are "universally recognized scientific achievements that for a time provide model problems and solutions to a community of practitioners" (*The Structure of Scientific Revolutions* [1962], p. x).
15. Basil, *Long Rules,* 48.
16. For Torvalds's description of his first programming experiments, such as a submarine game, see Learmonth, "Giving It All Away" (1997). Similarly, Wozniak became excited by technology in the fourth grade, and in the sixth he built a computer that played tic-tac-toe. Wozniak describes how his learning proceeded: "It was all self-done; I didn't even take a course, didn't ever buy a book on how

to do it" (Wolfson and Leyba, "Humble Hero"). In another context, he adds: "It's much more important to get a student motivated and want to learn something . . . than it is just strictly to teach it, teach it, teach it and expect that it gets absorbed" (Tech, "An Interview with Steve Wozniak" [1998]).

17. Torvalds, "Re: Writing an OS" (1992).
18. Cf. Plato, *Minos,* 319e.
19. Plato describes the idea of midwifery through the mouth of Socrates, whom he has say in one of his dialogues:

> one thing which I have in common with the ordinary midwives is that I myself am barren of wisdom. The common reproach against me is that I am always asking questions of other people but never express my own views about anything, because there is no wisdom in me; and that is true enough. And the reason of it is this, that God compels me to attend the travail of others, but has forbidden me to procreate. So that I am not in any sense a wise man; I cannot claim as the child of my own soul any discovery worth the name of wisdom. But with those who associate with me it is different. At first some of them may give the impression of being ignorant and stupid; but as time goes on and our association continues, all whom God permits are seen to make progress—a progress which is amazing both to other people and to themselves. And yet it is clear that this is not due to anything they have learned from me; it is that they discover within themselves a multitude of beautiful things, which they bring forth into the light. (Plato, *Theaetetus,* 150c–d)

Plutarch sums up: "Socrates was not engaged in teaching anything, but by exciting perplexities as if inducing the inception of labour-pains in young men he would arouse and quicken and help to deliver their innate conceptions; and his name for this was obstetric skill, since it does not, as other men pretend to do, implant in those who come upon it intelligence from without but shows that they have it native within themselves but undeveloped and confused and in need of nurture and stabilization" (*Platonic Questions,* 1000e).

The Socratic idea is that the purpose of teaching is to help someone learn to learn, to be able to pose questions. A precondition for that is puzzlement. In the dialogue *Meno,* the title character describes the Socratic teacher's effect:

> Socrates, before I even met you I used to hear that you are always in a state of perplexity and that you bring others to the same state, and now I think you are bewitching and beguiling me, simply putting me under a spell, so that I am quite perplexed. Indeed, if a joke is in order, you seem, in appearance and in every other way, to be like the electric ray, for it too makes anyone who comes close and touches it feel numb, and you now seem to have had that kind of effect on me, for both my mind and my tongue are numb, and I have no answer to give you. (80a–b)

But this state of perplexity is ultimately for the better, as Socrates explains:

> SOCRATES: Have we done him any harm by making him perplexed and numb as the torpedo fish does?
> MENO: I do not think so.
> SOCRATES: Indeed, we have probably achieved something relevant to finding out how matters stand, for now, as he does not know, he would be glad to find out, whereas before he thought he could easily make many fine speeches to large audiences about the square of double size and said that it must have a base twice as long.
> MENO: So it seems.
> SOCRATES: Do you think that before he would have tried to find out that which he thought he knew though he did not, before he fell into perplexity and realized he did not know and longed to know? (84a–c; see also *Alcibiades,* 106d)

20. The reason the Socratic teacher was also called a matchmaker was that it was his task to join people into giving birth together (Xenophon, *Symposium,* 3). Socrates describes his method: "With the best will in the world I undertake the business of match-making; and I think I am good enough—God willing—at guessing with whom they might profitably keep company. Many of them I have given away to Prodicus; and a great number also to other wise and inspired persons" (Plato, *Theaetetus,* 151b). Compare this to: "Someone asked Aristippus [a disciple of Socrates] how Socrates had helped him. He replied, 'He enabled me to find for myself satisfying fellow-students of philosophy' " (Philodemus, *Rhetoric,* 1, 342.13).

21. The third Academy metaphor was that of the teacher as master of ceremonies (the *symposiarkhos*) at banquets. These took place in the evenings, and in conjunction with the dialogues of the day they were an essential learning experience. The purpose of these banquets was quite serious and intellectually ambitious—for example, the discussion of some heavyweight philosophical subject—but they were, in addition, powerfully experiential events. (Two great descriptions are the symposia of Plato and Xenophon.)

    The symposiarch was responsible for the success of the banquets in two ways: first, from his elevated position he made sure that the intellectual goals of dialogue were attained; second, it was also his responsibility to ensure that none of the participants remained too stiff. To this latter end, he had two means at his disposal. First, he had the right to order the excessively stiff participants to drink more wine. If this did not work, the symposiarch could order the participant to remove his clothes and dance! The symposiarch used any means necessary to catalyze passionate contributions (cf. Plato, *Symposium,* 213e–14a).
22. Plato, *Republic,* 7.536e.
23. *The Rule of St. Benedict,* 6.
24. Slowly, these themes are winning more space in theories of education. There is a renewed interest in collaborative learning, inspired largely by Vygotsky's concept of the zone of proximal development, which stresses that an individual's potential capability when cooperating with a more experienced person is greater than his or her actual capability in isolation (*Mind in Society* [1978]). When learners set questions themselves and work together, they can also learn from each other—they can benefit from the fact that there are always some learners who are more advanced. This is why Lave and Wenger find it important for learners and researchers to be in dialogue with one another. They speak of the novice's "legitimate peripheral participation" in the expert culture (*Situated Learning: Legitimate Peripheral Participation* [1991]). The cautious formulation suggests what most university professors think about this idea.

Chapter 5: From Netiquette to a Nethic

1. The best expression of the netiquette that the hacker community shares is in "Netiquette Guidelines" by the Internet Engineering Task Force (RFC 1855), although it emphasizes that its purpose is not to "specify an Internet standard of any kind." Another important

expression of the netiquette is Vint Cerf's draft, "Guidelines for Conduct on and Use of Internet" (1994).

2. For the history of EFF, see Kapor and Barlow, "Across the Electronic Frontier" (1990), and Barlow, "A Not Terribly Brief History of the Electronic Frontier Foundation" (1990).
3. Barlow's most famous application of the word occurs in "A Declaration of the Independence of Cyberspace" (1996).
4. Cf. Ceruzzi, *History of Modern Computing* (1998), chaps. 8–9.
5. Gans and Goffman, "Mitch Kapor and John Barlow Interview" (1990).
6. Electronic Frontier Foundation, "About EFF."
7. The project is described in Electronic Frontier Foundation, *Cracking DES: Secrets of Encryption Research, Wiretap Politics, and Chip Design* (1998).
8. The Global Internet Liberty Campaign was formed at a meeting of the Internet Society to work for "prohibiting prior censorship of on-line communication" and "ensuring that personal information generated on the GII [Global Information Infrastructure] for one purpose is not used for an unrelated purpose or disclosed without the person's informed consent and enabling individuals to review personal information on the Internet and to correct inaccurate information," among other similar goals (cf. Global Internet Liberty Campaign, "Principles"). It links together the key organizations in both the freedom-of-expression and the privacy fields—e.g., the Center for Democracy and Technology (www.cdt.org), the Digital Freedom Network (www.dfn.org), the Electronic Frontier Foundation (www.eff.org), the Electronic Privacy Information Center (www.epic.org), the Internet Society (www.isoc.org), Privacy International (www.privacy.org/pi), and the XS4ALL Foundation (www.xs4all.net).

   Other important thematic alliances include the Internet Free Expression Alliance and the Internet Privacy Coalition.
9. For a global overview on the freedom of expression in cyberspace, see Dempsey and Weitzner, *Regardless of Frontiers: Protecting the Human Right to Freedom of Expression on the Global Internet;* Human Rights Watch, "Freedom of Expression on the Internet" (2000); and Sussman, *Censor Dot Gov: The Internet and Press Freedom 2000* (2000).
10. Sussman, *Censor Dot Gov* (2000), p. 1.
11. For general reports on the Kosovo war and the media, see Free

2000, *Restrictions on the Broadcast Media in FR Yugoslavia* (1998); Open Society Institute, *Censorship in Serbia;* Human Rights Watch, "Federal Republic of Yugoslavia," *World Report 2000* (2000); Reporters sans frontières, *Federal Republic of Yugoslavia: A State of Repression* and *War in Yugoslavia—Nato's Media Blunders.* For a more general commentary on the Kosovo war, with some references to information technology, see Ignatieff, *Virtual War: Kosovo and Beyond* (2000).

12. Cf. Joseph Saunders, *Deepening Authoritarianism in Serbia: The Purge of the Universities* (1999).
13. Open Society Institute, *Censorship in Serbia.*
14. The e-mails were published online by National Public Radio as "Letters from Kosovo" (1999).
15. Human Rights Watch, "Human Rights Defenders" and "Federal Republic of Yugoslavia," *World Report 2000* (2000); Committee to Protect Journalists, *Attacks on the Press in 1999*; and Reporters sans frontières, *Federal Republic of Yugoslavia.*
16. *Restrictions on the Broadcast Media,* pp. 16–17; XS4ALL, "The History of XS4ALL."
17. XS4ALL, "History of XS4ALL."
18. Human Rights Watch, "Federal Republic of Yugoslavia," *World Report 2000.*
19. Witness, *Witness Report 1998–1999.*
20. Witness, "About Witness" and *Witness Report 1998–1999.*
21. See OneWorld, "Internet to Play Major Role in Kosovo Refugee Crisis" (1999). The site was www.refugjat.org.
22. A personal communication from President Ahtisaari's assistant Matti Kalliokoski.
23. In addition to Denning's study, see Attrition.org, "Clinton and Hackers" (1999).
24. For some overviews on privacy in the information age, see Lessig, *Code and Other Laws of Cyberspace* (1999), chap. 11, and Gauntlett, *Net Spies: Who's Watching You on the Web?* (1999).
25. Human Rights Watch, "Freedom of Expression on the Internet."
26. Cf. Electronic Privacy Information Center, *Privacy and Human Rights 1999: An International Survey of Privacy Laws and Developments.*
27. *The New Hacker's Dictionary* (1996), appendix A, p. 514.
28. Gauntlett, *Net Spies,* p. 110.
29. For a review of the state of the regulation of cryptography in the

United States and elsewhere, see Madsen and Banisart, *Cryptography and Liberty 2000: An International Survey of Encryption Policy* (2000), and Koops, *Crypto Law Survey.*

30. Hughes, "A Cypherpunk's Manifesto" (1993).
31. Gilmore, "Privacy, Technology, and the Open Society" (1991). The third cofounder of the Cypherpunks, Tim May, has also written a manifesto, which he read at the group's founding meeting. Cf. "The Crypto Anarchist Manifesto" (1992).
32. Penet, "Johan Helsingius closes his Internet remailer" (1996) and Quittner, "Anonymously Yours—An Interview with Johan Helsingius" (1994). For a short history of Helsingius's anonymous remailer, see Helmers, "A Brief History of anon.penet.fi" (1997).
33. Baudrillard, *Amérique* (1986).
34. Andrew, *Closing the Iron Cage: The Scientific Management of Work and Leisure* (1981), p. 136.

Chapter 6: The Spirit of Informationalism

1. Castells, "Materials for an Exploratory Theory of the Network Society" (2000). "Self-programmable" workers correspond closely to what Reich calls "symbolic-analytic workers" in his *Work of Nations* (1991), chap. 14. The empirical data on the rise of this type of flexible work is provided by Carnoy, *Sustaining the New Economy* (2000), figs. 3.1–4. See also the study on work conditions in California—which, as the geographic center of information-technology development, often prefigures trends that are later witnessed elsewhere—by the University of California, San Francisco, and the Field Institute: according to it, two thirds of Californian workers are flex workers, and if we specify that only those who stay in their jobs for at least three years count as traditional workers, the figure rises to 78 percent (*The 1999 California Work and Health Survey* [1999]).
2. In his *Principles of Scientific Management* (1911), Taylor described the method for optimizing the motions of workers as follows:

> First. Find, say, 10 or 15 different men (preferably in as many separate establishments and different parts of country [as possible]) who are especially skillful in doing the particular work to be analyzed.
>
> Second. Study the exact series of elementary operations or motions which each of these men uses in doing the work

> which is being investigated, as well as the implements each man uses.
>
> Third. Study with a stop-watch the time required to make each of these elementary movements and then select the quickest way of doing each element of the work.
>
> Fourth. Eliminate all false movements, slow movements, and useless movements.
>
> Fifth. After doing away with all unnecessary movements, collect into one series the quickest and best movements as well as the best implements. (p. 61)

3. Robbins, *Awaken the Giant Within,* p. 274.
4. Franklin, *Autobiography,* p. 98.
5. Evagrius, 1, in Ward, *The Sayings of the Desert Fathers.* The whole passage reads:

> Imagine the fearful and terrible judgement. Consider the fate kept for sinners, their shame before the face of God and the angels and archangels and all men, that is to say, the punishments, the eternal fire, worms that rest not, the darkness, gnashing of teeth, fear and supplications. Consider also the good things in store for the righteous: confidence in the face of God the Father and His Son, the angels and archangels and all the people of the saints, the kingdom of heaven, and the gifts of that realm, joy and beatitude.
>
> Keep in mind the remembrance of these two realities. Weep for the judgement of sinners, afflict yourself for fear lest you too feel those pains. But rejoice and be glad at the lot of the righteous. Strive to obtain those joys but be a stranger to those pains. Whether you be inside or outside your cell, be careful that the remembrance of these things never leaves you, so that, thanks to their remembrance, you might at least flee wrong and harmful thoughts.

Compare this to Robbins: "People can succeed if they imagine something vividly enough just as easily as if they had the actual experiences" (*Awaken the Giant Within,* p. 80), and, "You see, ten years from now, you will surely arrive. The question is: Where? Who will you have become? How will you live?" (p. 31).

6. Robbins, *Awaken the Giant Within,* p. 31.

7. Franklin, *Autobiography*, pp. 86, 85.
8. Dorotheus, *Didaskaliai*, 104.1–3.
9. Athanasius, *Life of Anthony*, 55. Compare this to Robbins, who says that "the best strategy in almost any case is to find a role-model, someone who's already getting the results you want, and then to tap into their knowledge. Learn what they're doing, what their core beliefs are, and how they think" (*Awaken the Giant Within*, p. 25).
10. See especially Hadot's essays "Spiritual Exercises" and "Ancient Spiritual Exercises and 'Christian Philosophy,' " in *Philosophy as a Way of Life: Spiritual Exercises from Socrates to Foucault* (1995).
11. Robbins, *Awaken the Giant Within*, chap. 12.
12. Ibid., p. 44.
13. Franklin, *Autobiography*, p. 85.
14. Athanasius, *Life of Anthony*, 67.
15. Robbins, *Awaken the Giant Within*, pp. 216–18.
16. Franklin, *Autobiography*, p. 85.
17. Cassian, *Institutes of the Coenobia*, 9.4, 9.3.
18. Robbins, *Awaken the Giant Within*, p. 44.
19. Weber, *Protestant Ethic*, p. 53. Cf. Franklin, *Autobiography*, p. 81.
20. Cassian, *Institutes of the Coenobia*, 10.6.
21. Robbins, *Awaken the Giant Within*, p. 294. Cf. also Robbins's goal forms on pp. 277–80, 289–302.
22. Ibid., pp. 471–72. Here Robbins refers explicitly to Franklin's bookkeeping system.
23. Franklin, *Autobiography*, p. 59.
24. Ibid., p. 86.
25. Ibid., pp. 86–87.
26. Dorotheus, *Didaskaliai*, 111.13, 117.7.
27. Castells, *Information Age* (1996–1998), vol. 1, p. 199.
28. Ibid., vol. 3 (1998).
29. Weber, *Protestant Ethic*, p. 52.
30. *Stability Pact for South Eastern Europe* (1999).
31. Robbins, *Awaken the Giant Within*, p. 182.
32. Ibid., p. 27.
33. Internet Society, "Internet Society Guiding Principles."
34. Nua, *Internet Survey: How Many Online* (September 2000). According to this, there are about 380 million people online, of whom about 160 million are in the United States and Canada.
35. For more, see the NetDay webpage at www.netday.org.
36. Brand, *The Clock of the Long Now* (1999), pp. 2–3.

37. Danny Hillis, "The Millennium Clock."
38. The Mitchell Kapor Foundation, "The Mitchell Kapor Foundation Environmental Health Program."
39. Weeks, "Sandy Lerner, Network of One" (1998).

Chapter 7: Rest

1. Cited in Levy, *Hackers*, p. 236.
2. This is a question Augustine asks again and again. See *On Genesis Against the Manichees*, 1.2; *Confessions*, 11.13, 12; and *City of God*, 11.5. Augustine's own answer is that one cannot meaningfully speak of the time *before* Creation because creation does not take place within time and space but creates them as well.
3. Milton, *Paradise Lost* (1667).
4. Schneider, *The Other Life* (1920), p. 297.
5. When Dante descends to the Inferno in *Divine Comedy*, he meets Socrates, Plato, and other academics in the Limbo of Hell, continuing their dialogues (canto 4).
6. Genesis 1.2–4.

Appendix: A Brief History of Computer Hackerism

1. *The Gospel According to Tux.*
2. An instance of hacker humor, the acronymic name of the GNU enterprise to develop a Unix-like operating system and software was derived from the phrase "GNU's Not Unix." Stallman reacted against the closing of software source code, as exemplified by AT&T's decision to commercialize its Unix (which was developed in its Bell Labs). On October 27, 1983, Stallman sent a message to the newsgroups net.unix-wizards and net.usoft:

   > Free Unix!
   >
   > Starting this Thanksgiving I am going to write a complete Unix-compatible software system called GNU (for GNU's Not Unix), and give it away free to everyone who can use it. Contributions of time, money, programs and equipment are greatly needed.

   A little later, Stallman expanded this original message into an entire hacker statement of principles: "The GNU Manifesto" (1985). Stallman sees GNU as a spiritual successor to the open-source operating system designed by MIT's hackers as early as the late sixties, ITS (Incompatible Time-sharing System). The best-known creations of the GNU project are emacs, an editor favored by many hackers, and gcc, a translator of C-language, used by the Linux hackers.

For more on GNU's history, see Stallman, "The GNU Operating System and the Free Software Movement" (1999); for ITS, see Levy, *Hackers*, pp. 123–28.

3. The BSD project began in close cooperation with Bell Labs' Unix designers. When, in the early eighties, AT&T decided to commercialize the operating system, BSD became the nexus of the hackers' Unix development. In the nineties, BSD progressed along three main lines: NetBSD, FreeBSD, and OpenBSD. Details in Marshall McKusick, "Twenty Years of Berkeley Unix: From AT&T-Owned to Freely Redistributable" (1999).
4. While Thompson started the development of Unix, his collaboration with Ritchie, who developed the C-language to this end, was close from the very beginning. The histories of C-language and Unix thus have been closely intertwined. For more details on the history of Unix, see Ritchie, "The Evolution of the UNIX Time-Sharing System" and "Turing Award Lecture: Reflections on Software Research." See also Salus, *A Quarter Century of Unix* (1994).
5. For example, one often hears the claim that the aim of the Arpanet was to build a network impervious to nuclear attacks. In their essay "A Brief History of the Internet" (2000), the main movers in the Net's development (Vinton Cerf, Bob Kahn, et al.) have called this widespread belief a "false rumor." The Net's true origins were more practical. The project's director, Lawrence Roberts, an academic who moved from MIT to ARPA, envisaged a net as a means of advancing the cooperation of the computer scientists: "In particular fields of disciplines it will be possible to achieve a 'critical mass' of talent by allowing geographically separated people to work effectively in interaction with a system" (Roberts, "Multiple Computer Networks and Intercomputer Communication" [1967], p. 2).
6. The first Network Working Group was followed by the International Network Working Group (INWG), which was organized for the development of the Internet standards at the International Conference on Computer Communications in 1972. The working group's first director was Cerf. The INWG had no formal authority, but in practice it developed and established the Internet's most important standards (together with Bob Kahn, Cerf was central to the development

of the Internet's key protocols, TCP/IP (Transmission Control Protocol/Internet Protocol), which define how information is transmitted on the Net.

Finally, in the early eighties, ARPA officially retired from the Internet. After that, the central driving force in the development of the Net has increasingly been hackers. The INWG's successor, the Internet Engineering Task Force (IETF), was founded in 1986. It is completely open. In fact, the only way to be a "member" of this group is to participate in its open mailing-list discussion or meetings. Scott Bradner, one of the leading experts on the Internet infrastructure, sums up the role of this open group: "Apart from TCP/IP itself, all of the basic technology of the Internet was developed or has been refined in the IETF" ("The Internet Engineering Task Force" [1999], p. 47; for more on the IETF, see Bradner's article, Internet Engineering Task Force, "The Tao of IETF"; and Cerf, "IETF and ISOC"; for a brief description of the Internet Society, see its "All About the Internet Society").

When one considers the successfulness of the Internet's developmental model, it is worth remembering that TCP/IP was not the only suggestion of its time for a "network of networks." The two biggest standardization organizations, CCITT and OSI, had their own official standards (X.25 and ISO). On the basis of Abbate's research, it seems that one of the main reasons why these traditional standardization organizations' protocols did not succeed was the significantly more closed nature of these bodies' operation (*Inventing the Internet* [1999], chap. 5).

7. Abbate, *Inventing the Internet,* p. 127.
8. Berners-Lee, *Weaving the Web* (1999), p. 123. Berners-Lee was by no means the first to dream of a global hybertext. The best-known visionary of this idea is Ted Nelson, the inventor of the term *hypertext.* In his best-known work on the subject, *Literary Machines* (1981), Nelson for his part acknowledges his indebtedness to one of the most influential representatives of American information-processing technology, Vannevar Bush. As early as the nineteen-forties, Bush came up with the idea of a hypertext device he called Memex ("As We May Think" [1945]). Douglas Engelbart, active in the development of the Internet, presented his oNLine System as a product of his Augmenting Human Intellect research project in San Francisco in 1968: it contained many of the same elements now found in the Web. (For this demonstration, he also invented the mouse; cf. Ceruzzi, *A History of Modern Computing* [1998], p. 260;

for Engelbart's larger vision, see his "Augmenting Human Intellect: A Conceptual Framework" [1962]). In the humanities, the hypertext idea does, of course, have an even longer history (see, e.g., Landow, *Hypertext v.2.0* [1997]). Berners-Lee says, however, that he was not familiar with these visions when he developed his idea (p. 4).

At the time of its breakthrough, the Web had direct competitors, from which it differed to its advantage in its social model. Until 1994, the World Wide Web was essentially just one of many ideas for new utilizations of the Internet, and it was by no means clear which one of these would spearhead its evolution (nor was it even obvious that any of them would significantly influence the Internet). The most powerful competing idea was the Gopher information system developed by the University of Minnesota. Gopher hit the wall in the spring of 1993, when the decision was made to commercialize it. Berners-Lee describes this event: "This was an act of treason in the academic community and the Internet community. Even if the university never charged anyone a dime, the fact that the school had announced it was reserving the right to charge people for the use of the gopher protocols meant it had crossed the line" (p. 73). Berners-Lee made sure that CERN would allow him to keep the development of the Web entirely open (p. 74).

9. Berners-Lee, *Weaving the Web,* p. 47.
10. Michael Dertouzos, "Foreword," in ibid., p. x. One of the main goals of the World Wide Web Consortium (W3C) is to ensure the openness of the Web's key protocols (HTTP/URL [HyperText Transfer Protocol/Uniform Resource Locator] and HTML [HyperText Markup Language]), which define how webpages are transmitted over the Web and how their content is syntaxed. For more, see "About the World Wide Web Consortium."
11. For more on Andreessen's role in the development of the Web, cf. Robert H. Reid, *Architects of the Web: 1,000 Days That Built the Future of Business* (1997), chap. 1; John Naughton, *A Brief History of the Future: The Origins of the Internet* (1999), chap. 15; Berners-Lee, *Weaving the Web,* chap. 6. Andreessen went on to found Netscape with Jim Clark, who was at the time best known for being the founder of Silicon Graphics (cf. Clark, *Netscape Time*). Netscape closed the source code, which may have been its most fatal error in its lost fight with Microsoft's Internet Explorer (but there were also limits to the openness of Mosaic's source code that were set by the university's "Procedures for Licensing NCSA Mosaic" [1995]).

Netscape reissued its browser again as open source code in 1998 (called Mozilla), but it is uncertain if this helps anymore because the browser is already such a monster that it is very difficult for others to join in at this point (cf. "Mozilla.org: Our Mission" [2000]; Hamerly, Paquin, and Walton, "Freeing the Source: The Story of Mozilla" [1999]; Raymond, "The Revenge of the Hackers" [1999]).

The NCSA Web server, developed by student Rob McCool and others, had a similar explosive impact on the server side, as Mosaic had on the user side. (The user's browser is linked to the Web-server program at the server end.) McCool also joined Netscape. However, this part of the hacker heritage was saved more because the so-called Apache hackers, such as former Berkeley student Brian Behlendorf, started to develop the NCSA server further from the very beginning as open-source code.

Keith Porterfield summarizes the general dependence of the operation of the Internet and the Web on hacker creations by expressing what would happen in practice if the hacker programs were retracted from the technical core of them (my brief comments on the reasons are in parenthesis):

Over half the websites on the Internet would disappear (because about two thirds of the sites are run by them; cf. Netcraft, The Netcraft Web Server Survey [September 2000])
Usenet newsgroups would also go away (because they are supported by the hacker-created INN program)
But that wouldn't matter, because e-mail wouldn't be working (because most e-mail transmissions are made through the hacker-created Sendmail program)
You'll be typing "199.201.243.200" into your browser instead of "www.netaction.org" (because the Internet's plain-language "address list" depends on the hacker-created BIND program).

INN (InterNetNews) is the creation of hackers such as Rich Salz (see "INN: InterNetNews"). Sendmail was originally developed by a Berkeley student, Eric Allman, in 1979 (see "Sendmail.org"). BIND stands for Berkeley Internet Name Domain, and it was originally developed by Berkeley students Douglas Terry, Mark Painter, David Riggle, and Songnian Zhou (see "A Brief History of BIND" for other key people). All these hacker projects are presently carried on by the Internet Software Consortium (although its involvement in Sendmail takes place indirectly through its support of the Sendmail Consortium).

12. For details, see Campbell-Kelly and Aspray, *Computer: A History of the Information Machine* (1996), pp. 222–26, and Levy, *Hackers*, part 1.
13. Cf. Brand, "Fanatic Life and Symbolic Death Among the Computer Bums," in *II Cybernetic Frontiers;* Levy, *Hackers*, pp. 56–65. Later, this game led to the birth of the computer-game industry (cf. Herz, *Joystick Nation* [1997], chap. 1), whose sales figures are currently about the same as those of the movie industry in the United States (cf. Interactive Digital Software Association, *State of the Industry Report* [1999], p. 3).
14. Nelson, *Computer Lib*, introduction to the 1974 ed., p. 6. Cf. the jargon file, s.v. *cybercrud.* Through its predecessor, the People's Computer Company (which despite its name was not a business enterprise but rather a nonprofit organization), the group had connections to other parts of the sixties counterculture and favored its principle of giving power to the people. (Movements advancing freedom of speech, the status of women and homosexuals, the environment, and animals were strong in the Bay Area.) French and Fred Moore, the initiators of the Homebrew Computer Club, were both active in the PCC. They put this announcement on a notice board:

> AMATEUR COMPUTER USERS GROUP HOMEBREW COMPUTER CLUB . . . you name it
>
> Are you building your own computer? Terminal? TV Typewriter? I/O device? or some other digital black magic box?
>
> Or are you buying time on a time-sharing service?
>
> If so, you might like to come to a gathering of people with likeminded interests. Exchange information, swap ideas, help work on a project, whatever. (Levy, *Hackers*, p. 200)

PCC's founder, Bob Albrecht, promoted the use of computers in the fight against bureaucratic powers that be. The cover of the first issue of the PCC's journal (October 1972) carried this text: "Computers are mostly used against people instead of for people. Used to control people instead of to FREE them. Time to change all that—we need a People's Computer Company" (ibid., p. 172). One attendee at the

PCC's Wednesday-night meetings was Lee Felsenstein, a student at the University of California at Berkeley, who had also participated in the Free Speech Movement and the student occupation of a university building in December 1964. Felsenstein's goal was to provide people everywhere with the free use of computers. According to his proposal, this would provide "a communication system which allows people to make contact with each other on the basis of mutually expressed interests, without having to cede judgment to third parties" (ibid., p. 156). From the PCC group, both Albrecht and Felsenstein moved on to the Homebrew Computer Club, the latter acting as its discussion moderator at a later time.

15. Kennedy, "Steve Wozniak: Hacker and Humanitarian."
16. Ironically enough, Apple fell behind in its competition with the PC IBM launched in 1981 largely because, after its corporatization, Apple ended up with a closed architecture, in contrast to IBM (the old enemy of hackers), whose PC succeeded due to its open architecture, which made it possible for others to join in.

Bibliography

Abbate, Janet. *Inventing the Internet.* Cambridge, Mass.: MIT Press, 1999.

Andrew, Ed. *Closing the Iron Cage: The Scientific Management of Work and Leisure.* Montreal: Black Rose Books, 1999.

Anthony, Peter. *The Ideology of Work.* London: Tavistock, 1977.

Aristotle. *Politics.* In *The Complete Works of Aristotle: The Revised Oxford Translation.* Vol. 2. Trans. B. Jowett. Princeton: Princeton University Press, 1984.

Athanasius. *Life of Anthony.* In *Nicene and Post-Nicene Fathers.* 2d series, vol. 4. Trans. H. Ellershaw [1892]. Peabody, Mass.: Hendrickson, 1999.

Association for Democratic Initiatives. "About the Kosovar Refugee Database" (www.refugjat.org/aboutDbase.html).

Attrition.org "Clinton and Hackers." July 1999 (www.attrition.org/errata/art.0109.html).

Augustine. *Confessions.* Trans. R. S. Pine-Coffin. London: Penguin Books, 1961.

———. *Concerning the City of God Against the Pagans.* Trans. Henry Bettenson. London: Penguin Classics, 1972 (repr. 1984).

———. *On Genesis Against the Manichees.* In *The Fathers of the Church.* Vol. 84. Trans. R. J. Teske. Washington, D.C.: Catholic University of America Press, 1991.

Baltes, Matthias. "Plato's School, the Academy," *Hermathena* 140 (1993).

Barlow, John Perry. "A Not Terribly Brief History of the Electronic Frontier Foundation," 1990 (www.eff.org/pub/EFF/history.eff).

———. "A Declaration of the Independence of Cyberspace." Davos, 1996 (www.eff.org/~barlow/Declaration-Final.html).

Basil. *The Long Rules.* In *The Fathers of the Church.* Vol. 9. Trans. Sister M. Monica Wagner [1950]. Washington, D.C.: Catholic University of America Press, 1970.

Baudrillard. *Amérique.* Paris: Bernard Grasset, 1986.

Benedict. *The Rule of St. Benedict.* Trans. Boniface Atchison Verheyen. KA: St. Benedict's Abbey, 1949.

Berkeley Internet Name Domain. "A Brief History of BIND" (www.isc.org/products/BIND/bind-history.html).

Berners-Lee, Tim. *Weaving the Web: The Original Design and Ultimate Destiny of the World Wide Web by Its Inventor.* New York: HarperCollins, 1999.

Billot, M.-F. "Académie." In *Dictionnaire des philosophes antiques.* Ed. R. Goulet. Paris: Éditions du centre national de la recherche scientifique, 1989.

Borgman, Christine. *From Gutenberg to the Global Information Infrastructure: Access to Information in the Networked World.* Cambridge, Mass.: MIT Press, 2000.

Bradner, Scott. "The Internet Engineering Task Force." In DiBona, Ockham, and Stone, *Open Sources.*

Brand, Stewart. *II Cybernetic Frontiers.* New York and Berkeley: Random House and The Bookworks, 1974.

———. *The Media Lab: Inventing the Future at MIT.* New York: Viking, 1987.

———. *The Clock of the Long Now: Time and Responsibility.* New York: Basic Books, 1999.

Bunnell, David, with Adam Brate. *Making the Cisco Connection: The Story Behind the Real Internet Superpower.* New York: John Wiley and Sons, 2000.

Burton-Jones, Alan. *Knowledge Capitalism: Business, Work, and Learning in the New Economy.* Oxford: Oxford University Press, 1999.

Bush, Vannevar. "As We May Think." *Atlantic Monthly,* July 1945.

Cailliau, Robert. "A Little History of the World Wide Web." In World Wide Web Consortium, 1995 (www.w3.org/History.html).

Campbell-Kelly, Martin, and William Aspray. *Computer: A History of the Information Machine.* New York: Basic Books, 1996.

Capra, Fritjof. *The Web of Life*. New York: Random House, 1996.

Carnoy, Martin. *Sustaining the New Economy: Work, Family, and Community in the Information Age*. Cambridge, Mass.: Harvard University Press, 2000.

Cassian, John. *The Twelve Books on the Institutes of the Coenobia*. In *Nicene and Post-Nicene Fathers*. 2d series, vol. 11. Trans. Edgar Gibson [1894]. Peabody, Mass.: Hendrickson, 1999.

Castells, Manuel. *The Information Age: Economy, Society and Culture*. Vol. 1: *The Rise of the Network Society*. Malden, Mass.: Blackwell, 1996 (repr. 1997; 2d ed., 2000).

———. *The Information Age: Economy, Society and Culture*. Vol. 2: *The Power of Identity*. Malden, Mass.: Blackwell, 1997.

———. *The Information Age: Economy, Society and Culture*. Vol. 3: *End of Millennium*. Malden, Mass.: Blackwell, 1998 (2d ed., 2000).

———. "Materials for an Exploratory Theory of the Network Society." *British Journal of Sociology* 51:1 (2000).

Castells, Manuel, and Emma Kiselyova. *The Collapse of Soviet Communism: The View from the Information Society*. Berkeley: University of California International and Area Studies Book Series, 1995.

Cerf, Vinton. "Guidelines for Conduct on and Use of Internet" (draft). Reston, Va.: Internet Society, 1994 (www.isoc.org/internet/conduct/cerf-Aug-draft.shtml).

———. "IETF and ISOC," 1995 (www.isoc.org/internet/history/ietfhis.html).

Ceruzzi, Paul. *A History of Modern Computing*. Cambridge, Mass.: MIT Press, 2000.

Cherniss, H. F. *The Riddle of the Early Academy*. Berkeley and Los Angeles: University of California Press, 1945.

Clark, Jim, with Owen Edwards. *Netscape Time: The Making of the Billion-Dollar Start-Up That Took on Microsoft*. New York: St. Martin's Press, 1999.

Committee to Protect Journalists. *Attacks on the Press in 1999: A Worldwide Survey*. New York, 2000 (www.cpj.org/attacks99/frameset_att99/frameset_att99.html).

Connick. ". . . And Then There Was Apple," *Call-A.P.P.L.E.*, October 1986.

Copley, Frank. *Frederick W. Taylor: Father of the Scientific Management*. New York: Harper and Brothers, 1923.

Covey, Stephen. *The Seven Habits of Highly Effective People: Restoring the Character Ethic*. [1989.] New York: Simon and Schuster, 1999.

Crick, Francis. *The Astonishing Hypothesis.* New York: Charles Scribner's Sons, 1994.

Dante. *The Divine Comedy.* Trans. Mark Musa. New York: Penguin Books, 1984.

Davis, Stan, and Christopher Meyer. *Future Wealth.* Boston: Harvard Business School Press, 2000.

Defoe, Daniel. *Robinson Crusoe.* Ed. Angus Ross. London: Penguin Books, 1965 (repr. 1985).

Dell, Michael. *Direct from Dell: Strategies That Revolutionized an Industry.* London: HarperCollins Business, 2000.

Dempsey, Bert, Debra Weiss, Paul Jones, and Jane Greenberg. *A Quantitative Profile of a Community of Open Source Linux Developers.* Chapel Hill: School of Information and Library Science, University of North Carolina, 1999 (ils.unc.edu/ils/research/reports/TR-1999-05.pdf).

Dempsey, James, and Daniel Weitzner. *Regardless of Frontiers: Protecting the Human Right to Freedom of Expression on the Global Internet.* Global Internet Liberty Campaign (www.gilc.org/speech/report).

Denning, Dorothy, *Activism, Hacktivism, and Cyberterrorism: The Internet as a Tool for Influencing Foreign Policy.* Washington, D.C.: Georgetown University, 2000 (www.nautilus.org/info-policy/workshop/papers/denning.html).

DiBona, Chris, Sam Ockham, and Mark Stone, eds. *Open Sources: Voices from the Open Source Revolution.* Sebastopol, Calif.: O'Reilly and Associates, 1999 (www.oreilly.com/catalog/opensource/book/netrev.html).

Diffie, Whitfield, and Susan Landau. *Privacy on the Line: The Politics of Wiretapping and Encryption.* Cambridge, Mass.: MIT Press, 1999.

Dillon, John. "What Happened to Plato's Garden?" *Hermathena* 134 (1983).

Dusanic, S. "Plato's Academy and Timotheus' Policy, 365–359 B.C." *Chiron* 10 (1980).

Electronic Frontier Foundation. *Cracking DES: Secrets of Encryption Research, Wiretap Politics, and Chip Design.* San Francisco: Electronic Frontier Foundation, 1998.

———. "About EFF" (www.eff.org/abouteff.html).

Electronic Privacy Information Center. "Workplace Privacy." In *Privacy and Human Rights 2000: An International Survey of Privacy Laws and Developments* (www.privacyinternational.org/survey/phr2000/threats.html#Heading18).

Engelbart, Douglas. "Augmenting Human Intellect: A Conceptual Framework." Stanford: Stanford Research Institute, October 1962 (www.histech.rwth-aachen.de/www/quellen/engelbart/AHI62.pdf).

Epictetus. *Discourses.* Trans. W. A. Oldfather. Loeb Classical Library, vols. 131, 128 [1925, 1928]. Cambridge, Mass.: Harvard University Press, 1998, 1985.

Fischer, Claude. *America Calling: A Social History of the Telephone to 1940.* Berkeley and Los Angeles: University of California Press, 1992.

Flannery, Sarah, with David Flannery. *In Code: A Mathematical Journey.* London: Profile Books, 2000.

Franklin, Benjamin. "Advice to a Young Tradesman." In *The Writings of Benjamin Franklin,* vol. 2. Ed. Albert Henry. New York: Macmillan, 1905.

———. *Autobiography and Other Writings.* Ed. Ormond Seavey. Oxford: Oxford University Press, 1993 (reissued 1998).

FreeB92. "Keeping the Faith." April 1, 1999 (www.opennet.org/announcements/010499.shtml).

Free 2000. *Restrictions on the Broadcast Media.* September 1998 (www.free2000.opennet.org/pdf/publications.pdf).

Freiberger, Paul, and Michael Swaine. *Fire in the Valley: The Making of the Personal Computer.* 2d ed. New York: McGraw-Hill, 2000.

Gaiser, Konrad. *Philodems Academica: Die Bericht über Platon und die Alte Akademie in zwei herkulanensischen Papyri.* Stuttgart: Frommann-Holzboog, 1988.

Gans, David, and Ken Goffman. "Mitch Kapor and John Barlow Interview." *Wired,* August 1990 (www.eff.org/pub/Publications/John_Perry_Barlow/HTML/barlow_and_kapor_in_wired_interview.html).

Gardiner, Eileen, ed. *Medieval Visions of Heaven and Hell Before Dante.* New York: Italica Press, 1989.

———. *Medieval Visions of Heaven and Hell: A Sourcebook.* Garland Medieval Bibliographies, vol. 11. New York: Garland Publishing, 1993.

Gates, Bill. *The Road Ahead.* Rev. ed. New York: Penguin Books, 1996.

Gauntlett, Andrew. *Net Spies: Who's Watching You on the Web?* Berkeley: Frog, 1999.

Gilmore, John. "Privacy, Technology, and the Open Society." Speech given at the first conference on Computers, Freedom, and Privacy, March 28, 1991 (www.toad.com/gnu.cfp.talk.txt).

Global Internet Liberty Campaign. "Principles" (www.gilc.org/about/principles/html).

Glucker, John. *Antiochus and the Late Academy.* Göttingen: Vandenhoeck und Ruprecht, 1978.

Gold, Rebecca. *Steve Wozniak: A Wizard Called Woz.* Minneapolis: Lerner Publications, 1994.

Greenfield, Richard. *Censorship in Serbia.* New York: Open Society Institute, 1999 (www.soros.org/censorship/balkans/serbia.html).

Gregory the Great. *The Homilies of St. Gregory the Great on the Book of the Prophet Ezekiel.* Trans. T. Gray. Calif.: Etna, 1990.

Hadot, Pierre. "Spiritual Exercises." In *Philosophy as a Way of Life: Spiritual Exercises from Socrates to Foucault.* Trans. Michael Chase. Oxford: Blackwell, 1995.

———. "Ancient Spiritual Exercises and 'Christian Philosophy.' " In his *Philosophy as a Way of Life.*

Hafner, Katie, and Matthew Lyon. *Where Wizards Stay Up Late: The Origins of the Internet.* New York: Touchstone, 1998.

Hamerly, Jim, and Tom Paquin, with Susan Walton. "Freeing the Source: The Story of Mozilla," in DiBona, Ockman, and Stone, *Open Source.*

Hammer, Michael. "Reengineering: Don't Automate, Obliterate." *Harvard Business Review,* July–August 1990.

Hammer, Michael, and James Champy. *Reengineering the Corporation: A Manifesto for Business Revolution.* New York: HarperBusiness, 1994.

Hankins, J. "The Myth of the Platonic Academy of Florence." *Renaissance Quarterly* 44 (1991).

Held, David, Anthony McGrew, David Goldblatt, and Jonathan Perraton. *Global Transformations: Politics, Economics, and Culture.* Stanford: Stanford University Press, 1999.

Helmers, Sabine. "A Brief History of anon.penet.fi, the Legendary Anonymous Remailer." *Computer-Mediated Communication* Magazine 4:9 (1997) (December.com/cmc/mag/1997/sep/helmers.html).

Herz, J. C. *Joystick Nation: How Videogames Gobbled Our Money, Won Our Hearts, and Rewired Our Minds.* London: Abacus, 1997.

Hesiod. *Work and Days.* Trans. Hugh G. Evelyn-White. Cambridge, Mass.: Loeb Classical Library, Harvard University Press, 1914.

Hillis, Danny. "The Millennium Clock." *Wired,* 1995 (www.wired.com/wired/scenarios/clock.html).

Homer. *The Odyssey.* Trans. A. T. Murray, rev. George Dimock. Loeb

Classical Library, vols. 104–5, Cambridge, Mass.: Harvard University Press, 1995.

Hughes, Eric, "A Cypherpunk's Manifesto." March 9, 1993 (ftp://ftp.csua.berkeley.edu/pub/cypherpunks/rants/.manifesto.html).

Hughes, Thomas. *Rescuing Prometheus.* New York: Random House, 1998.

Human Rights Watch. *Human Rights Watch World Report 2000.* New York, 2000.

———. "Federal Republic of Yugoslavia." In *Human Rights Watch World Report 2000.*

———. "Freedom of Expression on the Internet." In *Human Rights Watch World Report 2000.*

———. "Human Rights Defenders." In *Human Rights Watch World Report 2000.*

Ignatieff, Michael. *Virtual War: Kosovo and Beyond.* New York: Metropolitan Books, 2000.

Interactive Digital Software Association. *State of the Industry Report,* 1999 (www.idsa.com/IDSA_SOTI_REPORT.pdf).

Internet Engineering Task Force. "The Tao of IETF," excerpted from RFC 1718 (www.ietf.cnri.reston.va.us/tao.html).

———. "Netiquette Guidelines." RFC 1855. (www.ietf.org/rfc/rfc1855.txt).

Internet Society. "All About the Internet Society" (www.isoc.org/isoc/).

———. "Internet Society Guiding Principles." (www.isoc.org/isoc/mission/principles).

Joy, Bill. "Why the Future Doesn't Need Us." *Wired,* April 2000 (www.wired.com/wired/archive/8.04.joy_pr.html).

Justin Martyr. *Apology.* In *Ante-Nicene Fathers.* Vol. 1. [1885.] Peabody, Mass.: Hendrickson Publishers, 1999.

Kantrowitz, Barbara, "Busy Around the Clock." *Newsweek,* July 17, 2000.

Kapor, Mitchell, and John Perry Barlow. "Across the Electronic Frontier," 1990 (www.eff.org/pub/EFF/electronic_frontier.eff).

Kennedy, John. "Steve Wozniak: Hacker and Humanitarian." In *Hindsights: The Wisdom and Breakthroughs of Remarkable People.* Ed. Guy Kawasaki. Beyond Words, 1994.

Koops, Bert-Jaap. *Crypto Law Survey* (cwis.kub.nl/~frw/people/koops/lawsurvy.htm).

Kuhn, Thomas. *The Structure of Scientific Revolutions.* Chicago: University of Chicago Press, 1962.

Landow, George. *Hypertext 2.0: The Convergence of Contemporary Criti-*

cal Theory and Technology. Baltimore: Johns Hopkins University Press, 1997.

Lavater, Johann Kasper. *Aussichten in die Ewigkeit.* Hamburg: Buchhandlergesellschaft, 1773.

Lave, J., and E. Wenger. *Situated Learning: Legitimate Peripheral Participation.* Cambridge: Cambridge University Press, 1991.

Learmonth, Michael. "Giving It All Away." *MetroActive,* May 8–14, 1997 (www.metroactive.com/papers/metro/05.08.97/cover/linus-9719.html).

Legion of Doom. "The History of the Legion of Doom." *Phrack* 31 (1990) (phrack.infonexus.com/search.phtml?view&article=p31-5).

Leiner, Barry, Vinton Cerf, David Clark, Robert Kahn, Leonard Kleinrock, Daniel Lynch, Jon Postel, Lawrence Roberts, and Stephen Wolff. "A Brief History of the Internet." Internet Society, 2000, (www.isoc.org/internet/history/brief.html).

Lennier. *Gospel of Tux,* 1999 (www.ao.com/~regan/penguins/tux.html).

Le Roy Ladurie, Emmanuel, *Montaillou: Cathars and Catholics in a French Village, 1294–1324.* Trans. Barbara Bray. London: Penguin Books, 1978.

Lesnick. *Preaching in Medieval Florence.* Athens, 1989.

Lessig, Lawrence. *Code and Other Laws of Cyberspace.* New York: Basic Books, 1999.

Levy, Steven. *Hackers: Heroes of the Computer Revolution.* New York: Delta, 1994.

Linzmayer, Owen. *Apple Confidential: The Real Story of Apple Computer, Inc.* San Francisco: No Starch Press, 1999.

Long Now Foundation. "Location" (www.longnow.org/10klibrary/Clock-Library_location.htm).

Lowe, Janet. *Bill Gates Speaks: Insight from the World's Greatest Entrepreneur.* New York: John Wiley and Sons, 1998.

Lyon, Jeff, and Peter Gorner. *Altered Fates: Gene Therapy and the Retooling of Human Life.* New York: W. W. Norton, 1995.

McKusick, Marshall Kirk. "Twenty Years of Berkeley Unix: From AT&T-Owned to Freely Redistributable." In DiBona, Ockman, and Stone, *Open Sources.*

Madsen, Wayne, and David Banisar. *Cryptography and Liberty 2000: An International Survey of Encryption Policy.* Washington, D.C.: Electronic Privacy Information Center, 2000 (www2.epic.org/reports/crypto2000).

Maslow, Abraham. *Motivation and Personality.* 3d ed. [1954.] New York: Longman, 3d ed., 1987.

———. *Toward a Psychology of Being.* 3d ed. [1962.] New York: John Wiley and Sons, 1999.

Matic, Veran. "Bombing the Baby with the Bathwater." March 30, 1999 (www.opennet.org/announcements/300399.shtml).

May, Tim. "The Crypto Anarchist Manifesto." 1992 (ftp://ftp.csua.berkeley.edu/pub/cypherpunks/rants/.crypto-anarchy.html).

Mentor. "The Conscience of a Hacker," *Phrack* 7 (1986) (phrack.infonexus.com/search.phtml?view&article=p7-3).

Merton, Robert. *The Sociology of Science: Theoretical and Empirical Investigations.* Ed. Norman Storer. Chicago: University of Chicago Press, 1973.

Microsoft. "Microsoft Timeline" (www.microsoft.com/billgates/bio).

Milton, John. *Paradise Lost.* Ed. Harold Bloom. New York: Chelsea House, 1996.

Mitchell Kapor Foundation. "The Mitchell Kapor Foundation Environmental Health Program" (www.mkf.org/envhlthmkf.html).

Mokyr, Joel. *The Lever of Riches: Technological Creativity and Economic Progress.* New York: Oxford University Press, 1990.

Moore, Gordon. "The Experts Look Ahead," *Electronics,* April 19, 1965.

Mozilla.org. "Mozilla.org: Our Mission." 2000 (www.mozilla.org/mission).

National Public Radio. "Letters from Kosovo." March 5–June 17, 1999 (npr.org/programs/morning/kosovo-emails.html).

Naughton, John. *A Brief History of the Future: The Origins of the Internet.* London: Weidenfeld and Nicolson, 1999.

Nelson, Ted. *Computer Lib/Dream Machines.* [1974.] Redmond, Wash.: Microsoft Press, 1987.

———. *Literary Machines: The Report on, and of, Project Xanadu Concerning Word Processing, Electronic Publishing, Hypertext, Thinkertoys, Tomorrow's Intellectual Revolution, and Certain Other Topics Including Knowledge, Education, and Freedom.* Self-published, 1981.

Netcraft. *The Netcraft Web Server Survey.* September 2000 (www.netcraft.com/survey/Reports/0009/).

Nua, *Internet Survey: How Many Online.* September 2000. (www.nya.ie/surveys/how_many_online/index.html).

Oikarinen, Jarkko. "Early IRC History." 1993 (www.irc.org/history_docs/jarkko.html).

OneWorld. "Internet to Play Major Role in Kosovo Refugee Crisis." April 9, 1999 (www.oneworld.org/about/ppack/releases/refugees_pr-rel.shtml).

Opensource.org. "History of the Open Source Initiative" (www.opensource.org/history.html).

Patterson, Robert. *Paradise: The Place and State of Saved Souls.* Philadelphia: Presbyterian Board of Publication, 1874.

Penet. "Johan Helsingius Closes His Internet Remailer." August 30, 1996 (www.penet.fi/press-english.html).

Perens, Bruce. "The Open Source Definition." In DiBona, Ockman, and Stone, *Open Sources,* and at www.opensource.org/osd.html.

Pine, Joseph, II, and James Gilmore. *The Experience Economy: Work Is Theatre and Every Business a Stage.* Boston: Harvard Business School Press, 1999.

Plato, *Alcibiades.* Trans. D. S. Hutchinson. In *Complete Works.*

———. *Apology.* In *Complete Works.* Ed. John M.Cooper with D. S. Hutchinson. Trans. G. M. A. Grube. Indianapolis: Hackett, 1997.

———. *Axiochus.* In *Complete Works.* Trans. Jackson P. Hershbell.

———. *Clitophon.* In *Complete Works.* Trans. Francisco J. Gonzales.

———. *Crito.* In *Complete Works.* Trans. G. M. A. Grube.

———. *Euthydemus.* In *Complete Works.* Trans. Rosamond Kent Sprague.

———. *Euthyphro.* In *Complete Works.* Trans. G. M. A. Grube.

———. *Gorgias.* In *Complete Works.* Trans. Donald Jeyl.

———. *Letters.* In *Complete Works.* Trans. Glen R. Morrow.

———. *Meno.* Trans. G. M. A. Grube. In *Complete Works.*

———. *Minos.* Trans. Malcolm Schofield. In *Complete Works.*

———. *Phaedrus.* In *Complete Works.* Trans. Alexander Nehamas and Paul Woodruff.

———. *Protagoras.* In *Complete Works.* Trans. Stanley Lombardo and Karen Bell.

———. *Republic.* In *Complete Works.* Trans. G. M. A. Grube and rev. C. D. C. Reeve.

———. *Symposium.* In *Complete Works.* Trans. Alexander Nehamas and Paul Woodruff.

———. *Theaetetus.* In *Complete Works.* Trans. M. J. Levett and rev. Myles Burnyeat.

Plutarch. *Platonic Questions.* Trans. Harold Cherniss. In *Moralia* 13, part 1. Cambridge, Mass.: Loeb Classical Library, Harvard University Press, 1976.

Porterfield, Keith W. "Information Wants to Be Valuable." NetAction (www.netaction.org/articles/freesoft.html).

Quittner, Joshua. "Anonymously Yours—An Interview with Johan

Helsingius." *Wired* 2.06 (June 1994) (www.wired.com/wired/2.06/departments/electrosphere/anonymous.1.html).

Raymond, Eric. "A Brief History of Hackerdom." In DiBona, Ockman, and Stone, *Open Sources,* and www.tuxedo.org/~esr.writings/cathedral-bazaar/hacker-history/ (first version 1992).

———. "How to Become a Hacker." In Raymond, *Cathedral and the Bazaar* (www.tuxedo.org/~esr/faqs/hacker-howto.html) (first version 1996).

———. *The Cathedral and the Bazaar: Musings on Linux and Open Source by an Accidental Revolutionary.* Sebastopol, Calif.: O'Reilly and Associates, 1999 (www.tuxedo.org/~esr/writings/homesteading/cathedral-bazaar/) (first version 1997).

———. "Homesteading the Noosphere." In Raymond, *Cathedral and the Bazaar* (www.tuxedo.org/~esr/writings/homesteading/homesteading) (first version 1998).

———. "The Revenge of the Hackers." In Raymond, *Cathedral and the Bazaar* (www.tuxedo.org/~esr/writings/homesteading/hacker-revenge) (first version 1999).

———. "The Art of Unix Programming." 2000. Draft.

Raymond, Eric, ed. *The Jargon File,* 2000 (www.tuxedo.org/~esr/jargon).

———. *The New Hacker's Dictionary.* 3d ed. Cambridge, Mass.: MIT Press, 1998.

Reich, Robert. *The Work of Nations: Preparing Ourselves for Twenty-first-century Capitalism.* New York: Vintage Books, 1992 [1991].

Reid, Robert. *Architects of the Web: 1,000 Days That Built the Future of Business.* New York: John Wiley and Sons, 1997.

Reporters sans frontières. *Federal Republic of Yugoslavia: A State of Repression.* 1999 (www.rsf.fr/uk/rapport/yougo/rapportyougo.html).

———. *1999 Survey* (www.rsf.fr/uk/cp.protest/bilan99.html).

———. *War in Yugoslavia: Nato's Media Blunders.* 1999 (www.rsf.fr/uk/rapport/nato/nato.html).

Rheingold, Howard. *Tools for Thought: The History and Future of Mind-Expanding Technology.* Cambridge, Mass.: MIT Press, 2000.

Rifkin, Jeremy. *The End of Work: The Decline of the Global Labor Force and the Dawn of the Post-Market Era.* New York: G. P. Putnam's Sons, 1995.

Ritchie, Dennis. "The Evolution of the UNIX Time-Sharing System." *AT&T Bell Laboratories Technical Journal* 63:8 (1984).

———. "Turing Award Lecture: Reflections on Software Research." *Communications of the ACM* 27:8 (1984).

Robbins, Anthony. *Awaken the Giant Within: How to Take Immediate*

Control of Your Mental, Emotional, Physical, and Financial Destiny! New York: Fireside, 1992.

Roberts. "Multiple Computer Networks and Intercomputer Communication." Proceedings of ACM Symposium on Operating System Principles, Gatlinburg, Tenn., 1967.

Rosenberg, Donald. *Open Source: The Unauthorized White Papers.* Foster City, Calif.: IDG Books, 2000.

Russell Hochschild, Arlie. *The Time Bind: When Work Becomes Home and Home Becomes Work.* New York: Metropolitan Books, 1997.

Rybczynski, Witold. *Waiting for the Weekend.* New York: Penguin Books, 1992.

Salus, Peter. *A Quarter Century of UNIX.* Reading, Mass.: Addison-Wesley, 1994.

Saunders, Joseph. *Deepening Authoritarianism in Serbia: The Purge of the Universities. Human Rights Watch Short Report* 11:2 (1999).

Schneider. *The Other Life.* Rev. and ed. Herbert Thurston. New York: Wagner, 1920.

Sendmail.org. "Sendmail.org" (www.sendmail.org).

Smith, Adam. *Wealth of Nations.* [1776.] Oxford: Oxford University Press, 1993.

Solomon, Alan. "A Brief History of PC Viruses." S&S International, 1990 (www.bocklabs.wisc.edu/~janda/solomhis.html).

Southwick, Karen. *High Noon: The Inside Story of Scott McNealy and the Rise of Sun Microsystems.* New York: John Wiley and Sons, 1999.

Spector, Robert. *Amazon.com: Get Big Fast (Inside the Revolutionary Business Model That Changed the World).* London: Random House Business Books, 2000.

Stability Pact for South Eastern Europe. Cologne, June 10, 1999 (www.seerecon.org/KeyDocuments/KD1999062401.htm).

Stallman, Richard. "The GNU Manifesto." 1993 (first version 1985) (www.gnu.org/gnu/manifesto.html).

———. "What Is Free Software?" 2000 (first version 1996) (www.gnu.org/philosophy/free-sw.html).

———. "The GNU Operating System and the Free Software Movement." In DiBona, Ockman, and Stone, *Open Sources,* and www.gnu.org/gnu/thegnuproject.

———. "The Free Software Song" (www.org/music/free-software-song.html).

Sun Microsystems. "SUN Microsystems Co-Founder Resigns." August 8, 1995 (www.sun.com/smi/Press/sunflash/9508/sunflash.950810.3737.html).

Sussman, Leonard. *Censor Dot Gov: The Internet and Press Freedom 2000.* Freedom House, 2000 (www.freedomhouse.org/pfs2000/pfs2000.pdf).

Tanenbaum, Andrew. *Operating Systems: Design and Implementation.* Englewood Cliffs, N.J.: Prentice-Hall, 1987.

Taylor, Frederick Winslow. *The Principles of Scientific Management.* [1911.] Mineola, N.Y.: Dover Publications, 1998.

Tech. "An Interview with Steve Wozniak." October 26, 1998 (www.thetech.org/people/interviews/woz.html).

Tertullian. *The Prescription Against Heretics.* Trans. Peter Holmes. In *Ante-Nicene Fathers,* vol. 3. [1885.] Peabody, Mass.: Hendrickson Publishers, 1999.

Thompson, Edward. *The Making of the English Working Class.* [1963.] New York: Penguin Books, 1991.

———. "Time, Work-Discipline, and Industrial Capitalism." *Past and Present* 38 (1967).

Torvalds, Linus. "What Would You Like to See Most in Minix?" Message to comp.os.minix, August 25, 1991.

———. "Free Minix-like Kernel Source for 386-AT." Message to comp.os.minix, October 5, 1991.

———. "Re: Writing an OS." Message to linux-activists@bloom-picayune.mit.edu, May 5, 1992.

———. "Birthday." Message to linux-activists@bloom-picayune.mit.edu, July 31, 1992.

———. "Credits" (ftp://ftp.kernel.org/pub/linux/kernel/CREDITS).

Tournier, Michel. *Friday.* [1967.] Trans. Norman Denny. Baltimore: Johns Hopkins University Press, 1997.

Tuomi, Ilkka. *Corporate Knowledge. Theory and Practice of Intelligent Organizations.* Helsinki: Metaxis, 1999.

Ulyat, William Clarke. *The First Years of the Life of the Redeemed After Death.* New York: Abbey Press, 1901.

University of California, San Francisco, and the Field Institute. *The 1999 California Work and Health Survey.* 1999.

University of Illinois, "Procedures for Licensing NCSA Mosaic." July 19, 1995 (www.ncsa.uiuc.edu/SDG/Software/Mosaic/License/LicenseInfo.html).

Valloppillil, Vinod. *Open Source Software.* Microsoft Confidential, August 11, 1998 (www.opensource.org/halloween/halloween1.html).

Valloppillil, Vinod, and Josh Cohen. *Linux OS Competitive Analysis.* Microsoft Confidential, August 11, 1998 (www.opensource.org/halloween/halloween2.html).

van den Hoven, Birgit. *Work in Ancient and Medieval Thought: Ancient Philosophers, Medieval Monks and Theologians and Their Concept of Work, Occupations, and Technology.* Leiden, 1996.

Vygotsky, L. S. *Mind in Society: The Development of Higher Psychological Processes.* Cambridge, Mass.: Harvard University Press, 1978.

Ward, Benedicta, ed. *The Sayings of the Desert Fathers.* 1975.

Watts, Duncan. *Small World: The Dynamics of Networks between Order and Randomness* Princeton: Princeton University Press, 1999.

Wayner, Peter. *Free for All: How Linux and the Free Software Movement Undercut the High-Tech Titans.* New York: HarperBusiness, 2000.

Weber, Max. *The Protestant Ethic and the Spirit of Capitalism.* [1904–1905; trans. 1930.] London: Routledge, 1992.

Weeks, Linton. "Sandy Lerner, Network of One." *The Washington Post.* March 25, 1998 (www.washingtonpost.com/wp-srv/frompost/march98/lerner25.htm).

Wells, Joe. "Virus Timeline." IBM Antivirus Online, 1996 (www.bock-abs.wisc.edu/[illegible]).

Witness. *Witness Report 1998–1999* (witness.org/about/report9899.htm).

———. "About Witness" (witness.org/about.htm).

Wolfson, Jill, and John Leyba. "Humble Hero." San Jose Mercury Center (www.mercurycenter.com/archives/revolutionaries/wozniak.htm).

World Wide Web Consortium. "About the World Wide Web Consortium" (www.w3.org/Consortium).

Xenophon. *Symposium.* Trans. O. J. Todd. [1923.] Loeb Classical Library, vol. 168. Cambridge, Mass.: Harvard University Press, 1997.

XS4ALL. "The History of XS4ALL" (www.xs4all.net/uk/absoluut/history/index_e.html).

Young, Robert, and Wendy Goldman Rohm. *Under the Radar: How Red Hat Changed the Software Business—and Took Microsoft by Surprise.* Scottsdale, Ariz.: Coriolis, 1999.

Yutang, Lin. *The Importance of Living.* [1938.] Stockholm: Zephyr Books, 1944.

Acknowledgments

The writing of this book has been animated by its topic: the hacker ethic. The first thing was not a decision to write a book; the first thing was a belief in a certain way of life, and the book is just one of its outcomes. Life during the writing of this book—both while literally writing and while doing other things—has certainly been passionate and freely rhythmed. Sometimes this has meant stopping at the big questions and playing with thoughts without any hurry; sometimes this has meant a more hectic period of concentrated effort, which is necessary in even essentially playful action, as hackers have emphasized.

It has been a joy to work on this book with Linus and Manuel and to find that our themes are very close to one another's. I thank them and their families for all the great time spent together. I also want to thank warmly the many other special people with whom I have had a chance to work, especially my friend Henning Gutmann, who has lived in the spirit strongly and helped in so many ways, and the wonderful team of Scott Moyers, Timothy Mennel, and Sunshine Lucas, among others at Random House, that has shown what cooperation between the people at a publisher and an author can be at its best.

And finally, I would like to express my gratitude to my loved ones, for your being you—that inspires me in life tremendously.

ABOUT THE AUTHORS

PEKKA HIMANEN earned his Ph.D. in philosophy from the University of Helsinki at the age of twenty. His ongoing mapping of the meaning of technological development has brought him into dialogue with academics, artists, government ministers, and CEOs. Himanen works at the University of Helsinki and at the University of California at Berkeley.

LINUS TORVALDS has become one of the most respected hackers within the computer community for creating the Linux operating system in 1991 while a student at the University of Helsinki. Since then, Linux has grown into a project involving thousands of programmers and millions of users worldwide.

MANUEL CASTELLS is a professor of sociology at the University of California at Berkeley. He is the author of the acclaimed trilogy *The Information Age* and of *The City and the Grassroots* (winner of the 1983 C. Wright Mills Award) and of more than twenty other books.

ABOUT THE TYPE

This book was set in Bodoni Book, a typeface named after Giambattista Bodoni, an Italian printer and type designer of the late eighteenth and early nineteenth centuries. It is not actually one of Bodoni's fonts but a modern version based on his style and manner and is distinguished by a marked contrast between the thick and thin elements of the letters.